Design of Transmission Systems

Design of Transmission Systems

Erik Carver

CLANRYE
INTERNATIONAL
www.clanryeinternational.com

Clanrye International,
750 Third Avenue, 9th Floor,
New York, NY 10017, USA

ISBN: 978-1-64726-645-5

Cataloging-in-Publication Data

Design of transmission systems / Erik Carver.
    p. cm.
Includes bibliographical references and index.
ISBN 978-1-64726-645-5
1. Telecommunication systems. 2. Wireless power transmission. 3. Electric power transmission.
4. Telecommunication. 5. Electric power systems. I. Carver, Erik.
TK5101 .D47 2023
621.382--dc23

For information on all Clanrye International publications
visit our website at www.clanryeinternational.com

Contents

Preface

Over the recent decade, advancements and applications have progressed exponentially. This has led to the increased interest in this field and projects are being conducted to enhance knowledge. The main objective of this book is to present some of the critical challenges and provide insights into possible solutions. This book will answer the varied questions that arise in the field and also provide an increased scope for furthering studies.

Electric power transmission refers to the bulk movement of electrical energy from the site of generation such as power plant to an electrical substation. There are different tools that are used in transmission system design. These tools are transmission route identification and selection, transmission network expansion planning, network analysis, and reliability analysis. In order to analyze the performance of a specific transmission system, a system planner uses tools such as load-flow, stability, and short-circuit programs. Automatic expansion models are also sometimes used to determine the optimum system. The automatic expansion models can be categorized into three types including heuristic models, single-stage optimization models, and time-phased optimization models. This book explores the design of transmission systems in detail. It presents this complex subject in the most comprehensible and easy to understand language. The book will serve as a valuable source of reference for graduate and postgraduate students.

I hope that this book, with its visionary approach, will be a valuable addition and will promote interest among readers. Each of the authors has provided their extraordinary competence in their specific fields by providing different perspectives as they come from diverse nations and regions. I thank them for their contributions.

Erik Carver

Design of Flexible Elements

1.1   Design of Flat Belts and Pulleys

A flat belts drives can be used for large amount of power transmission and there is no upper limit of distance between the two pulleys. Belt conveyers system is one such example. These drives are efficient at high speeds and they offer quite running. A typical flat belt drive with idler pulley is shown in Figure given. Idler pulleys are used to guide a flat belt in various manners, but do not contribute to power transmission.

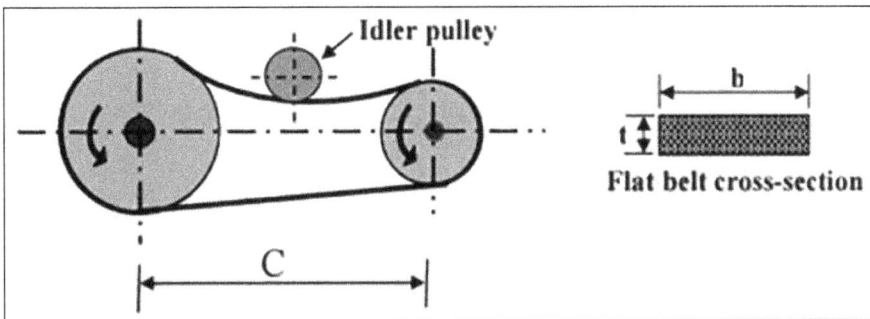

Belt Drive with Idler.

The flat belts are marketed in the form of coils. Flat belts are available for a wide range of width, thickness, weight and material. The depending upon the requirement one has to cut the required belt length from the coil and join the ends together. The fixing of the joint must be done properly because the belt normally gets snapped from the improper joints. The best way is to use a cemented belt from the factory itself or with care one can join these belts with various types of clips that are available in the market.

Types of Belts used for Transmission of Power

- Flat Belts.
- V-Belts.
- Ribbed Belts.
- Toothed or Timing Belts.

Materials used for Making Belts

- Leather.

- Cotton and Canvas.

- Rubber.

- Nylon core belts.

- Balata Belts.

Advantages

- A V-belt drive provides compactness due to the small distance between centers of pulleys.

- The drive is positive, as a result of the slip between the belt and the pulley groove is negligible.

- Since the V-belts are made endless and there is no joint trouble, therefore the drive is smooth.

- It provides longer life, three to 5 years.

- It can be easily installed and removed.

- The operation of the belt and machine is quiet.

- The belts have the ability to cushion the shock when machines are started.

- The high velocity ratio (maximum 10) is also obtained.

- The wedging action of the belt in the groove gives high value of limiting *ratio of tensions. Therefore the power transmitted by V-belts is more than flat belts for the same coefficient of friction, arc of contact and allowable tension in the belts.

- The V-belt may be operated in either direction, with tight side of the belt at the highest or bottom. The center line may be horizontal, vertical or inclined.

Disadvantages

- The V-belt drive cannot be used with large center distances, because of larger weight per unit length.

- The V-belts are not as durable as flat belts.

- The development of pulleys for V-belts is more complicated than pulleys of flat belts.

- Since the V-belts are subjected to certain amount of creep, therefore these are not suitable for constant speed applications like synchronous machines and timing devices.

- The belt life is greatly influenced with temperature changes, improper belt tension and mismatching of belt lengths.

- The centrifugal tension prevents the use of V-belts at speeds below five m/s and above 50 m/s.

In order to design a flat belt drive using manufacturer's data, you need to find ten different parameters. They are:

- Pulley Diameters (D and d).

- Speeds of Driving and Driven Pulleys (N_1 and N_2).

- Design Power in KW.

- Velocity of belt (V m/s).

- Selection of belt.

- Number of plies.

- Load rating at V m/s.

- Belt width (b).

- Pulley width.

- Length of belt (L).

Problems

1. Design a flat belts drive to transmit 6 kW at 900 rpm of the driver pulley. Speed reduction is to be 2.5.1. Assume that the service is 16 hours a day.

Given data:

- P = 6 kw

- N_1 = 900 rpm

- i = 2.5

Solution

1. Assume center distance C = 500 mm:

$$\frac{C}{D} = 1$$

$$\frac{500}{D} = 1$$

$$D = \frac{500}{1} = 500 \text{ mm}$$

$$D = 500 \text{ mm}$$

$$\frac{D}{d} = 2.5$$

$$\frac{500}{2.5} = d$$

$$d = 200 \text{ mm}$$

2. Calculation of design power in kW:

$$\text{Design kW} = \frac{\text{Rated kW} \times \text{Load Correction Factor}\left(k_d\right)}{\text{Acc. of Contact Factor}\left(k_a\right) \times \text{Small Pulley Factor}\left(k_d\right)}$$

Rated kW=6kW

Load correction factor k_s = 1.2 for steady load.

$$\text{Arc of Contact} = 180° - \left(\frac{D-d}{c}\right) \times 60°$$

$$= 180° - \left(\frac{500-200}{500}\right) \times 60°$$

$$= 144°$$

Arc of contact for 144°, kα = 1.13

Small pulley factor, kd = 0.7

$$\text{Design kw} = \frac{6 \times 1.2}{1.3 \times 0.7}$$

$$= 7.9 \text{ kW}$$

3. Selection of belt: Hi-Speed duck bating is selected. Its capacity is given as 0.023 kw/mm/ply.

4. Load rating correction:

$$\text{Velocity of the Bat } V = \frac{\pi d N_1}{60} = \frac{\pi \times 0.12 \times 900}{60}$$
$$= 9.42 \, m/e$$

Load rating at rm/s = Load rating at 10 n.s × V/10

Load rating at 9.42 m/s = 0.023 × (9.42/10)

$$= 0.02 \text{ kw/mm/ply.}$$

5. Determination of best width:

For 200 mm pulley diameter and velocity 9.42 m/s. The number of pulleys can be selected as 5.

$$\text{Width of Bat} = \frac{\text{Design Power}}{\text{Load Rating} \times \text{Number of Plies}}$$
$$= \frac{7.9}{0.02 \times 5} = 79 \text{ mm}$$

Per 5 ply salt, standard bat width = 76 mm.

6. Determination of pulley width:

Pulley width = Best width of 13 mm

$$= 76 + 13$$
$$= 89 \text{ mm}$$

Standard pulley width is 90 mm.

7. Calculation of length of the bat (α):

$$\alpha = 2c + \pi/2(D+d) + \frac{(D-d)^2}{4c}$$
$$= 2 \times 500 + \pi/2(500+200) + \frac{(500r\,200)^2}{4 \times 500}$$
$$= 2099.57 \text{ mm.}$$

2. Select a flat belt to drive a mill at 250 rpm from a 10 kW, 730 rpm motor. Center distance is to be around 2 m. The mill shaft pulley is of 1 m diameter.

Solution

Given data:

- Driver motor speed = 730 rpm.
- Power = 10 kW.
- Mill driven pulley speed = 250 rpm.
- Center distance = 2 m.
- Mill shaft pulley diameter = 1 m.

Find: Design the flat bolt drive.

1. Find out the driven, driver pulley diameters:

Consider motor pulley as driver pulley:

- $D_1 D_2$.
- D_1 – Driver pulley diameter = D_1.
- D_2 – Driven pulley diameter = 1m = 1000 mm.
- M_1 – Driver pulley speed = 730 rpm.
- M_2 – Driven pulley speed = 250 rpm.

$$D_1 = \frac{N_2 D_2}{N_1} = \frac{250 \times 1000}{730} = 342 \text{ mm}$$
$$N_1 D_1 = N_2 D_2$$
$$D_1 = 34.2 \text{ mm}$$

2. Calculate design power:

$$\text{Design Power} = \frac{\text{Rated Power i} \cdot (\text{kw}) \times k_{\text{Service}}}{k_{\text{arc factor}} \times k_{\text{small power factor}}}$$

Service factor = k_{sec} = 1.2 for steady load.

3. To get Arc Factor:

$$\text{Arc of contact} = 180° - \left(\frac{D-1}{C} \times 6\right)$$

$$= 180° - \left(\frac{D_2 - D_1}{C} \times 60°\right)$$

$$= 180° - \left(\frac{1000 - 342}{2000}\right) \times 60°$$

$$= 180° - 19.74°$$

$\theta = 160.24°$ which is more than $160°$, so select convection factors from the data book corresponding to the 170, $K_{\text{arc factor}} = 1.04$.

4. To get small pulley factor:

From the data book CP responding to the smaller pulley diameter of 342 mm diameter select the factor:

$$K_{\text{smaller pulley factor}} = 0.8$$

So,

$$K_{\text{service}} = 1\ 2$$

$$K_{\text{arc factor}} = 1.04$$

$$K_{\text{small, pulley factor}} = 0.8$$

So, the design power $= \dfrac{100 \times 1.2}{1.04 \times 0.8} = 14.4 \text{ m} = 14.5 \text{ kw.}$

5. Selection of belt:

Fabric belt, high speed luck belt is chosen, it's capacity is 0.023 kw/mm/ply from the PSG Design Data Book.

6. Load rating correction:

$$\text{Speed of the belt} = \frac{\pi D_1 \cdot N_1}{60} = \frac{\pi \times 342 \times 750}{60}$$

$$= 13.42 \text{ m / sec.}$$

$$\text{Load rating at } 13.42 \text{ m / s} = \text{Load rating at } 10 \text{ m / s} \times (13.4/10)$$

$$= (0.023 \times 13.42)/10$$

$$= 0.03087 \text{ kw / mm / ply}$$

7. Determination of belt width:

$$\text{Width of Belt} = \frac{\text{Design Power}}{\text{Load Rating} \times \text{Number of Plies}}$$

$$= \frac{14.5}{0.03037 \times 4}$$

$$= 117 \text{ mm}$$

8. Calculate the belt length:

$$\text{Open Belt Drive} = L = 2C + \frac{\pi}{2}\left(D_1 + D_2\right) + \frac{\left(D_1 - r\right)^2}{4C}$$

$$= 2 \times 2000 + \frac{\pi}{2}\left(342 + 1000\right) + \frac{\left(342 - 1000\right)^2}{4 \times 2000}$$

$$= 6061 \text{ mm}$$

Length of the flat open back Drive = 6161 mm.

$$\text{Pulley Width} = \text{Belt Width} + 13 \text{ mm}$$

$$= 117 \text{ mm} + 13 \text{ mm}$$

$$= 130 \text{ mm}$$

Result:

Belt width = 117 mm.

Belt length = 6161 mm = 6.161 m.

Width of pulley = 130 mm.

- D_1 = 342 mm
- D_2 = 1000 mm
- N_1 = 730 rpm
- N_2 = 250 rpm
- Power = 10 kw

3. A flat belt drive is to design to drive a flour mill. The driving power requirement of the mill is 22.5 kW at 750 rpm with a speed reduction of 3.0. The distance between the shafts is 3 m. Diameter of the mill pulley is 1.2 m. Let us a design and draw the sketch of the drive.

Given: Application - Flow Mill

- P = 22.5 kW

- N1 = 750 rpm

- i = 3

- C = 3 m

- D = 1.2 m

Solution

Step 1:

Given: $i = 3 = \dfrac{D}{d}$

$$D = 1.2 \text{ m}$$
$$\Rightarrow d = 0.4 \text{ m}$$

Step 2:

Design Power = Rated power × Load correction factor × Arc of contact factor.

Load Correction Factor = 1.5.

$$\text{Area of Contact, } \theta = 180 - \left(\frac{D-d}{C}\right)60°$$
$$\theta = 164°$$

Arc of contact factor = 1.06

∴ Design power = 35.775 kW

Step 3:

Dunlop "FORT" 949 of fabric belting for shock loads.

Step 4:

$$V = \frac{\pi d N_1}{60} = 15.7 \text{m} / \text{s}$$

Load rating at 10 m/s = 0.0289 kW/mm ply.

$$\text{Load rating at V m} / \text{s} = 0.0289 \times \frac{15.7}{10}$$
$$= 0.0453 \text{ kW} / \text{mm ply}$$

Step 5:

$$\text{Belt Width} = \frac{\text{Design Power}}{\text{Load Rating} \times \text{Number of Piles}}$$

Number of plies = 6

∴ Belt width = 131.62 mm

Standard belt width = 152 mm

Step 6:

Pulley width = 152 + 25 = 177 mm

Step 7:

Assume open drive:

$$L = 2C + \frac{\pi}{2}(D+d) + \frac{(D-d)^2}{4C}$$
$$= 8566.6 \text{ mm}$$

4. A compressor is to run by a motor pulley running at 1440 rpm, speed ratio 2.5. Choose a flat belt crossed drive. Center distance between pulleys is 3.6 m. Take belt speed as 16 m/s. Load factor is 1.3. Take a 5-ply, flat Dunlop belt. Power to be transmitted is 12 kW. High speed load rating is 0.0118 kW/ply/mm width of V = 5 m/s. Let us determine the width and length of the belt.

Given data:

- N_1 = 1440 rpm

- Speed ratio = 2.5

- C = 3.6 m

- Belt speed = 16 m/s

- L factor = 1.3

- Number of Ply = 5

- P = 12 kW = 12 × 10³ W

- High speed load rating = 0.0118 kW/Ply/mm

- $V = 5$ m/s

- Belt speed is 16 m/s

- $d = 250$ mm

Solution

1. Speed Ratio:

$$\text{Speed Ratio} = \frac{D}{d}$$

$$2.5 = \frac{D}{250}$$

$$D = 2.5 \times 250 = 625 \text{ mm.}$$

2. Calculation of design power in kW:

$$\text{Design kW} = \frac{\text{Rated kW} \times \text{Load Correction Factor}\left(k_\delta\right)}{\text{Arc of Contact Factor}\left(k_\alpha\right) \times \text{Small Pulley Factor}\left(k_\Omega\right)}.$$

- Rated kW = 12 Kw

- Load factor, $_{k\delta}$ = 1.3 (given)

To find arc of contact factor (kd):

$$\text{Arc of Contact} = 180° - \left(\frac{D-d}{C}\right) \times 60°$$

$$= 180° - \left(\frac{625-250}{3600}\right) \times 60°$$

$$= 173.75°$$

Contact area = 173.54°

$$k\alpha = 1.03$$

Smaller pulley diameter is 250.

So, kd = 0.7

$$\text{Design kW} = \frac{12 \times 1.3}{1.03 \times 0.7} = 21.637 \text{ kW}$$

3. Select of Belt: HI-SPEED duck belting is selected.

4. Load rating correction:

$$\text{Load Rating at V m/s} = \text{Load Rating and 10m/s} \times \frac{V}{10}$$

$$= 0.0118 \times \frac{16}{10} = 0.01888 \text{ kW/mm/kg}$$

5. Determination of belt width:

$$\text{Width of Belt} = \frac{\text{Design Power}}{\text{Loading Rating} \times \text{Number of Piles}}$$

$$= \frac{21.637}{0.01888 \times 5} = 229.205 \text{ mm}$$

Standard belt width = 250 mm

6. Calculation of length of the belt (L):

For crossed belt drive:

$$L = 2C + \frac{\pi}{2}(D+d) + \frac{(D+d)^2}{4C}$$

$$L = 2 \times 3600 + \frac{\pi}{2}(625+250) + \frac{(625+250)^2}{4 \times 3600}$$

$$= 7200 + 1374.45 + 53.168$$

$$= 8627.6 \text{ mm}$$

1.2   Selection of V Belts and Pulleys

Open Belt

An open belt drive is used to rotate the driven pulley in the same direction of driving pulley. In motion of belt drive, power transmission results makes one side of pulley more tightened compared to the other side. In horizontal drives, tightened side is always kept in the lower side of two pulleys because the sag of the upper side slightly increases the angle of folding of the belt on the two pulleys.

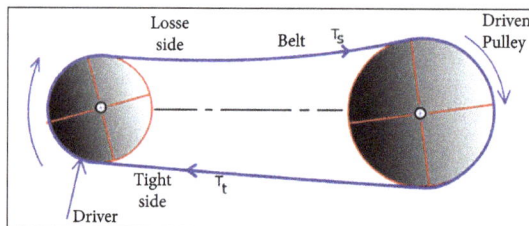

Open Belt Drive.

Cross Belt

A crossed belt drive is employed to rotate driven block within the wrong way of driving block. Higher the worth of wrap allows additional powers are often transmitted than associate open belt drive. However, bending and wear of the belt are vital considerations.

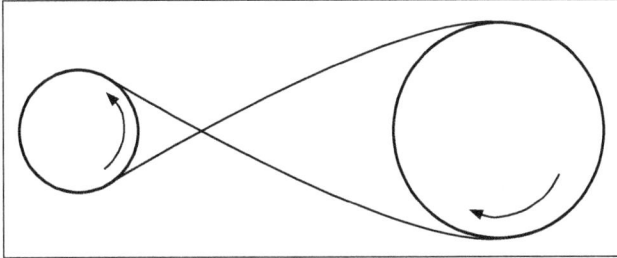

Crossed Belt Drive.

Difference between Open Drive and Cross Drive of a Belt Drive

Open Belt Drive	Cross Belt Drive
Open belt drive is used with shaft arranged parallel and rotating in same direction.	Crossed belt drive is used with shafts arranged parallel and rotating in opposite direction.
Length of belt: $$L_C = \frac{\pi}{2}(d_1 + d_2) + 2x + \frac{(d_1 - d_2)^2}{4x}$$	Length of belt: $$L_C = \frac{\pi}{2}(d_1 + d_2) + 2x + \frac{(d_1 - d_2)^2}{4x}$$

Advantages of V-Belt

- Smooth starting and running.

- Permit a wide range of driven speeds, using standard electric motors.

- They are rugged and provide years of trouble-free performance with minimal attention even under adverse conditions.

- Capable of transmitting power around corners or out of plane drives.

- Clean require no lubrication.

- Highly efficient.

- Extremely wide horsepower ranges.

- Dampen vibration between driver and driven machines.

- Silent operation.

- Long service life.

- Easy installation.

- Can be used as an effective means of clutching.

Design Procedure of V-Belt

Step 1: Selection of belt section PSG 7.58 -

- Select the cross section of belt based on power to be transmitted.

Step 2: Selection of pulley diameter PSG 7.58 -

- Select if not in the question.

- Select larger diameter of pulley for the given speed ratio.

Step 3: Selection of centre distance C PSG 7.61.

Step 4: Determination of nominal pitch length L PSG 7.61 -

- V-belts are designated by cross section & inside length.

- Standard inside length from PSG 7.58, 7.59, 7.60.

Step 5: Calculation of design power -

Design power = Rated power × service factor PSG7.69.

Arc of contact factor (PSG7.68) × correction factor length.

Step 6: Determination of maximum power capacity of V-belt PSG 7.62.

Step 7: Determination of no. of belt -

No. of belt = Design power/Rating of belt.

Step 8: Recalculation of centre distance PSG 7.61 -

$$C = A + \sqrt{A^2 - B}$$
$$A = L/4 - \pi(D+d)/8$$
$$B = (D-d)^2/8$$

Step 9: Calculate the details of V-groove pulley PSG 7.70.

Step 10: Write the specifications of the drive.

Problems

1. Let us design a V-belt drive to drive a machine at 400 rpm from a motor running at 1440 rpm. Assume suitable parameters.

Given:

$$N_1 = 1440 \text{ rpm and } N_2 = 400 \text{ rpm}$$

Find: Design a V-belt drive

Solution

1. Selection of the belt section: Assumed for power 7.5 kW, B section is selected.

2. Selection of pulley diameter (d and D).

For B section, recommended pulley diameter, d = 125 mm

$$\text{Speed Ratio} = \frac{D}{d} = \frac{N_1}{N_2} = \frac{1440}{400} = 36$$

Large pulley diameter, D = 36 d

$$D = 450 \text{ mm.}$$

3. Selection of center distance:

$$\text{For } i = \frac{D}{d} = 3.6$$

C/D ratio is 0.95.

$$\frac{C}{D} = 0.95$$

$$C = 427.5 \text{ mm}$$

4. Determination of nominal pitch length:

$$L = 2C + \frac{\pi}{2}(D+d) + \frac{(D-d)^2}{4C}$$

$$= 2 \times 427.5 + \frac{\pi}{2}(450+125) + \frac{(450+125)^2}{4 \times 427.5}$$

$$= 1819.9 \text{ mm}$$

The next standard pitch length is 1948 mm.

5. Selection of various modification factors:

Length Correction Factor (FC): For B Section Length correction factor, $F_c = 0.97$.

Correction factor for arc of contact (F_d):

$$\text{Arc of Contact} = 180 - \left(\frac{D-d}{C}\right) \times 60$$

$$= 180 - \left(\frac{450-125}{1948}\right) \times 60$$

$$= 169.9°$$

For this arc of contact, $F_d = 0.98$.

- Service Factor:

Light duty 16 hour continued service, of driving units-II.

$F_a = 1.3$

6. Calculation of maximum power capacity, for B section:

$$kW = \left(0.79\ S^{-0.09} - \frac{50.8}{de} - 1.32 \times 10^{-4}\ S^2\right)S$$

$$S = \text{Belt Speed} = \frac{\pi d N_1}{60} = \frac{\pi \times 0.125 \times 1440}{60} = 9.42\,m/s.$$

$$\text{Power, } kW = \left(0.79 \times 9.42^{-0.09} - \frac{50.8}{175} - 1.32 \times 10^{-4} \times 9.42^2\right)9.42$$

$$= 3.24\ kW$$

7. Determination of number of belts (nb):

$$n_b = \frac{P \times F_a}{kW \times F_c \times F_d}$$

$$= \frac{7.5 \times 1.3}{3.24 \times 0.97 \times 0.98}$$

$$= 3.6 = 4\ \text{Belts,}\left(\text{Assume power } P = 7.5\ kW\right).$$

8. Calculation of actual center distance:

$$C_{actual} = A + \sqrt{A^2 + B}$$

$$A = \frac{L}{4} - \pi\left[\frac{D+d}{8}\right] = \frac{1948}{4} - \pi\left[\frac{450+125}{8}\right] = 261.1$$

$$B = \frac{(D-d)^2}{8} = \frac{(450-125)^2}{8} = 13203.125$$

$$C_{actual} = 546.36 \text{ mm.}$$

2. A V-belt drive is to transmit 45 kW in a heavy duty saw mill which works in two shifts of 8 hours each. The speed of motor shaft is 1400 rpm with the approximate speed reduction of 3 in the machine shaft. Let us Design the drive and calculate the average stress induced in the belt.

Solution

Given: P = 45 kW; Heavy duty saw mill, 2 shifts of 8 hours each, n = 1400 rpm, i = 3.

For P = 45 kW, the cross section is 'C' d = 250 mm.

$$i = \frac{n_1}{n_2} = \frac{D}{d}$$

$$S = \frac{1400}{n_2} = \frac{D}{250}$$

$$D = 750 \text{ mm}$$

For $1 = 3\dfrac{C}{D} = 1$

$$C = 750 \text{ mm}$$

$$C_{min} = 0.55\,(D+d) + T = 0.55(750+250) + 14 = 564 \text{ mm}$$

$$C_{max} = 2(D+d) = 2(750+250) = 2000 \text{ mm}$$

Nominal inside length:

$$npL = 2C + \frac{\pi}{2}(D+d) + \frac{(D-d)^2}{4C}$$

$$= 3154 \text{ mm}$$

Std npL = 3205 mm

$\therefore$ Nominal inside length, Nil = 3150 mm.

Power rating:

$$kW = \left(1.47S^{-0.09} - \frac{142.7}{d_e} - 2.34 \times 10^{-4}S^2\right)S$$

$$S = \frac{\pi d n_1}{60} = 18.3 \text{ m/s}$$

For $\dfrac{D}{d} = 3$, $F_b = 1.14$

$(\therefore de = dp \times Fb = 285 \text{ mm})$

$\Rightarrow kW = 10.11$.

Standard KW rating = 9.41 kW

$$\text{Number of belts} = \frac{P \times F_a}{kW \times F_c \times F_d}$$

$F_a = 1.4$ (sawmill).

$$\theta = 180° - 60°\left(\frac{D-d}{C}\right) = 140°$$

$F_d = 0.9$

$F_c = 0.97$

$\therefore$ Number of belts = 8.

3. A V-belt drive consists of three V-belts in parallel in grooved pulleys of the same size. The angle of groove is 30° and the coefficients of friction 0.12. The cross sectional area of each belt is 800mm2 and the permissible safe stress in the belt material is 3 M Pa. Let us calculate the power that can be transmitted between two pulleys 400mm in diameter rotating at 960 rpm.

Given Data:

$\theta = 30°$

$= 30 \times \dfrac{\pi}{180}$

$\alpha = 0.523$ rad/s

$\mu = 0.12$

$A = 800 \text{ mm}^2$

$\Sigma \text{ belt} = 3 \text{ M pa}$

$N = 960 \text{ rpm}$

$P = 400 \text{ mm}$

$$\frac{T_1}{T_2} = e^{\mu\alpha}$$

$$\frac{T_1}{T_2} = e^{(0.12 \times 0.523)}$$

$$\frac{T_1}{T_2} = 1.004$$

Solution:

$$V = \frac{\pi DN}{60}$$

$$= \frac{\pi \times 400 \times 10^{-3} \times 960}{60}$$

$T_1 = 1.064 \, T_2$

$V = 20.1 \text{ m/s}$

Since, the velocity of the bolt is more than 10 m/s. Therefore the centrifugal tension must be taken in to consideration. Assuming a leather belt for which the density is given as 1000 kg/m3.

$T_c = MV_2$

Maximum tension in the belt:

$T = \sigma \text{ belt} \times \text{Area of cross section of the belt} = 3 \times 106 \times (b \times t)$

$T = 3 \times 106 \times 800 \times 10^{-6}$

$T = 2400 \text{ N}$

$T_c = M \times V_2$

Mass = Density × Volume

M = Density × Area × Length

$= 1000 \times 800 \times 10{-6} \times 1$

$M = 0.8$

$T_C = 0.8 \times 20.1$

$T_C = 16.08$

Tension on tight side of the belt:

$T_1 = T - T_C$

$T_1 = 2400 - 16.08$

$T_1 = 2383.92 \ N$

$T_2 = \dfrac{2383.92}{1.064}$

$T_2 = 2240.5 \ N$

$P = (T_1 - T_2)V$

$\quad = (2383.92 - 2240.5) \times 20.1$

$P = 288.7 \ W = \dfrac{2882.7}{3} = 960.9 \ W$

$P = 2.8 \ kW.$

The power carried per belt = 0.96 kW

4. A motor of power 2 kW running at a speed of 1400 rpm transmits power to an air blower running at 560 rpm. The motor pulley diameter is 200 mm. The center distance may be 1000 mm. Let us design a suitable V-belt drive.

Given:

$P = 2 \ kW$

$N_1 = 1400 \ rpm$

Application:

Air Blower:

$N_2 = 560 \ rpm$

$d = 200 \ mm$

$C = 1000 \ mm$

Solution:

Step 1:

$$\therefore \frac{N_1}{N_2} = \frac{D}{d}$$

d = 200 mm.

$$\Rightarrow D = 200 \times \frac{1400}{560} = 500 \text{ mm}$$

Step 2:

For p = 2 kW, cross section is "A".

Step 3:

C = 1000 mm (given).

Step 4:

Nominal pitch length (NPL):

$$L = 2C + \frac{\pi}{2}(D+d) + \frac{(D-d)^2}{4C}$$

L = 3122 mm

NPL_{std} = 3084 mm

∴ Corresponding NIL = 3048 mm

Step 5:

Power rating:

$$kW = \left(0.455^{-0.09} - \frac{19.62}{d_e} - 0.765 \times 10^{-4} S^2 \right) S.$$

$$S = \frac{\pi d N_1}{60} = 14.6 \, \text{m} / \text{s}$$

d_e = 200 × 1.13 = 226 mm

kW = 3.6

kWstd = 2.65

Step 6:

$$\text{Number of belts, } n = \frac{P \times F_a}{kW \times F_c \times F_d}$$

Assume Fa = 1

Fc = 1.13

$$\text{Area of Contact}, \theta = 180° - 60°\left(\frac{500-200}{1000}\right) = 162°$$

$\therefore$ Fd = 0.96

$$\therefore n = \frac{2 \times 1}{3.65 \times 1.13 \times 0.96} = 1$$

Step 7:

Actual center distance:

$$C = A + \sqrt{A^2 - B}$$

$$A = \frac{L}{4} - \pi\left(\frac{D+d}{8}\right) = 496.1$$

$$B = \frac{(D-d)^2}{8} = 11250$$

$$\Rightarrow C = 980.72 \text{ mm}$$

Step 8:

$$A - \frac{3048}{120} - 65$$

1.3   Selection of Hoisting Wire Ropes and Pulleys

These pulley blocks are used as an anti-friction device to lift heavy object. Construction industries, automobile and machinery industries are our main customers as we offer non-corrosive and durable products. We manufacture these blocks from high-grade stainless steel sourced from trustworthy vendors of the market. These Wire Rope Pulley Blocks have very high weight lifting capacity.

Various Stresses Induced in the Wire Ropes

- Direct stress.
- Bending stress.
- Stress due to acceleration.
- Stress during starting and stopping.
- Effective stress.

Crowning of Pulley

When the belt passes from slack side to tight side, certain portion of belt extends and it contracts again when the belt passes from tight side to slack side. One to these changes in length, there is a relative motion between belt and pulley called creep.

Problems

1. Let us a select wire-rope for a vertical mine hoist to lift a load of 30 kN from a depth of 600 m. A rope speed of 3 m/s is to be attained in 10 seconds.

Given:

$h = 600$ m

$W = 30$ kN

Velocity $= 3$ m/s $= 180$ m/min

$t = 10$ sec

To design wire rope.

Solution

1. For elevator 6 × 19 rope is selected.

2. Calculation of design load:

Design load = Load to be lifted × assumed FOS.

$= 30 \times 10 = 300$ kN

3. Selection of wire rope diameter (d):

$d = 25$ mm for $\sigma_u = 1600 - 1750$ N/mm² breaking strength $= 340$ kN

4. Calculation of sheave diameter:

$$6 \times 19 \text{ and Class } 4 \frac{D_{min}}{d} = 27$$

Lifting speed of 180 m/min for every additional speed ratio has to be increased 8%.

$$\frac{D_{min}}{d} = 27 \times (1.08)^4 = 36.73 \text{ say } 40$$

Sleeve diameter D = 40 × d = 1000 m

$$\text{Area A} = 0.4 \times \frac{\pi}{4} \times 25^2 = 196.35 \text{ mm}^2$$

6. Wire diameter:

$$d_N = \frac{d}{1.5\sqrt{i}} = \frac{25}{1.5\sqrt{114}} = 1.56 \text{ mm}$$
$$i = 6 \times 19 = 114$$

7. Weight of rope Wr:

Approximate weight = 2.41 kg f/m

$$= 24.1 \text{ N/m}$$

$$W_r = 24.1 \times 600$$

$$= 14.4 \text{ kN}$$

8. Various loads:

i. Acceleration load:

$$W_a = \left(\frac{W + W_r}{g}\right) a$$

$$a = \frac{V_2 - V_1}{t} = \frac{\frac{180}{60} - 0}{10} = 0.3 \text{ m/s}^2$$

$$= \left(\frac{30 + 14.4}{9.81}\right) 0.3 = 1.35 \text{ kN}$$

ii. Bending load:

$$W_b = \sigma_b \times A = E \times \frac{d_w}{D} \times A$$

$$= 0.84 \times 10^5 \times \frac{1.56}{1000} \times 196.35$$

$$= 25.729 \text{ kN}$$

iii. Direct load W + Wr = 30 + 14.4 = 44.4 kN

Effective load:

Wsbea = Wd + Wb + Wa

$\qquad$ = 44.4 + 25.72 + 1.35

$\qquad$ = 71.47 kN

9. Factor of safety:

$$F_{sw} = \frac{Breaking\,Stress}{\omega_{ae}} = \frac{300 \times 10^3}{71.47 \times 10^3} = 4.19$$

$\qquad$ n = 5

$\qquad$ FSN < n.

The design is not safe.

Redesign the same procedure by changing breaking stress value.

2. A workshop crane carries a load of 30 kN using wire ropes and a hook. The hook weighs 15 kN. Diameter of the rope drum is 30 times the diameter of the rope. The load is lifted with an acceleration of 1 m/s². Let us determine the diameter of the rope. FS = 6, E_r = 80 kN/mm2, σ_u = 180 kN/mm², cross section of the rope = 0.4 × (diameter of rope)².

Given Data:

$\qquad$ Load = 30 kN

$\qquad$ Hook Weight = 15 kN

Diameters of loop drum 30. × Diameter rope

$\qquad$ Acceleration = 1 m/s²

$\qquad$ F_{os} = 6

$\qquad$ E_r = 80 kN/mm²

$\qquad$ σ_U = 180 kN/mm²

Cross-section of rope = 0.4 (dia. rope)²

Solution

1. Assume a liberal factor of safety of 15 to obtain the design load.

Design load = 15 × dead weight on rope.

Deal weight = Load + Book Weight

$$= 30kN + 15\ kN$$

$$= 45\ kN$$

So, design load = 15 × 45

$$= 675\ kN$$

2. Since the rope is for hoisting purpose we use 6 × 19 rope; with reference to PSG design data book page number 9.6. Diameter = 25 mm, for tensile strength of 1100 N/mm to 1250 N/mm2 and breaking strength = 230 kN.

3. Shear diameter, from PSG design data book page no. 9.1 for velocity upto 50 m/min, $\dfrac{D_{min}}{d} = 30$

$$D = 30 \times d$$

$$= 30 \times 25$$

$$D = 750\ mm$$

$$\frac{D_{min}}{d} = 30$$

$$D_{min} = 30 \times d$$

$$= 30 \times 25$$

$$D_{min} = 750\ mm$$

Area of useful cross section of rope:

$$A = 0.Ad^2 \text{ for } 6 \times 19 \text{ rope}$$

$$= 0.4 \times 25^2$$

$$A = 250\ mm^2$$

Wire diameter:

$$d_w = \frac{d}{1.5\sqrt{i}} = \frac{25}{1.5\sqrt{6 \times 19}} = 1.56\ mm.$$

Weight of rope/motor = 2.25 × 9.81 = 22.5 N/m

Calculation of other loads:

Direct load, $W_d = W + W_{hook}$

$$= 30,000 + 15,000$$

$$= 45,000 \text{ N}$$

Bending Load, $W_b = \sigma_b \times A = E_4 \cdot \dfrac{d_w}{D} A$

$$= \frac{80 \times 10^3 \times 1.56 \times 250}{750}$$

$$= 41600 \text{ N}$$

Effective Load on the rope during normal:

Working condition $= W_{ne} = \omega_d + \omega_b$

$$= 45000 \text{ N} + 41600 \text{ N}$$

$$W_{ne} = 86600 \text{ N}$$

Load due to the acceleration $= W_a = \left(\dfrac{W + W_r}{g} \right) a.$

$$W_a = \frac{45000}{9.81} \times 1$$

$$= 4587 \text{ N}$$

Effective load during acceleration:

$$W_{ac} = W_{a+} W_d + W_b,$$

$$W_{ac} = 458 + 45000 + 41600 = 91187 \text{ N}$$

Factor of Safety and Number of Ropes:

Working Factor of Safety $= FS_W = \dfrac{\text{Breaking Strength}}{W_{ae}}$

$$= \frac{230000}{91187} = 2.5$$

From the table, we find the recommended factor of safety for this application to be n = 6. To achieve this factor of safety we may either chosen rope and drum with larger diameters (or) three two ropes.

$$\text{Number of Rates} = \frac{n}{FS_W} = \frac{6}{2.5} = 2.4 \ \Omega \ 3.$$

$$\begin{aligned} \text{Stress} &= \frac{\text{Total Load}}{\text{Areas}} \\ &= \frac{91107}{250} = 364 \ N/mm^2. \end{aligned}$$

Result:

Diameter of wire $= d_w = 1.56$ mm.

Number of rope = 3.

3. The construction site, 1 tonne of steel is to be lifted up to a height of 20 m width the help of 2 wire ropes of 6×19 size, nominal diameter 12 mm, and breaking load 78 kN. Let us determine the factor of safety if the sheave diameter is 56 d and if wire rope is suddenly stopped in 1 second when traveling at a speed of 1.2 m/s. Also, find the factor of safety, if bending load is neglected.

Given data:

Height = 20 m; W = 1 tonne = 1000 × 9.81 ω = 9810 N

Nominal diameter = 12 mm

Breaking load 78 kN = 78 × 10³ N, V=1.2 m/s

1. Selection of suitable wire rope, 6 × 19 rope [given].

2. Calculation of design load:

Design load = Load to be lifted × Assumed factor of safety

= 98 × 15 = 147.15 kN

3. Selection of wire rope diameter:

d = 18 mm for σ_u = 1600 to 1750 N/mm² and breaking strength = 193 kN

4. Calculation of sheave diameter (D):

$$\frac{D_{min}}{d} = 27 \ \text{(for velocity up to 50 m/min.)}$$

Since the given lifting speed 1.2 m/s, therefore Dmin/d ratio should be modified. Thus, for every additional speed of 50 m/min. $\dfrac{D_{min}}{d}$ ratio has to be increased by 8%.

$$\text{Modified } \frac{D_{min}}{d} = 27 \times (1.08)^{1.5-1}$$
$$= 28.059 \text{ say } 30$$

The sheave diameter, D = 3.0 × d = 30 × 18 = 540 mm

5. Selection of the area of useful cross-section of the rope (A): for 6 × 19 rope

A = 0.4012 = 0.4 × (18)2 = 130 mm²

6. Calculation of wire diameter (dw):

$$\text{Wire diameter}, d_w = \frac{d}{1.5\sqrt{i}}$$

i = Number of Strand × Number of wires in each Strand

= 6 × 19 = 114

$$d_W = \frac{18}{1.5\sqrt{114}} = 1.124 \text{ mm}$$

7. Selection of weight of rope (Wr):

Approximate mass = 2.41 kg/m

Weight of rope/m = 2.41 × 9.81 = 23.6 N/m

Weight of rope, W_r = 23.6 × 20 = 472 N

8. Calculation of various loads:

Direct load, W_d = W + W_r = 9810 + 472 = 10282 N

Bonding load:

$$W_b = \sigma_b \times A = \frac{E_r \cdot d_w}{D} \times A = \frac{0.84 \times 10^5 \times 1.124}{540} \times 130.$$

$$W_b = 22729.78 \text{ N}$$

Take E = 0.84 × 10⁵ N /mm²

Acceleration load:

$$W_a = \left(\frac{W + W_r}{9}\right) \cdot a = \left(\frac{9810 + 472}{9.81}\right) \times 1.2$$

W_a = 1257.73 N

Acceleration of the load, a = $\dfrac{V_2 - V_1}{t_1} = \dfrac{1.2 - 0}{1} = 1.2 \ m/s^2$

$$W_a = 1257.73 \ N$$

Starting load (W_{sf}):

$$w_{st} = 2 \cdot w_d = 2 \times (W + w_r) = 2 \times (9810 + 412)$$

W_{st} = 2056.4 N

9. Calculation of effective loads on the rope:

Effective load during normal working:

$$w_{en} = w_d + w_b$$

W_{en} = 10282 + 22729.78 = 33011.78 N

Effective load during acceleration of the load:

$$W_{ea} = W_d + W_b + W_a$$

$$= 10282 + 22729.78 + 1257.73 = 34269.51 \ N$$

Effective load during starting:

$$W_{est} = W_b + W_{st}$$

W_{est} = 22729.78 + 20564 = 43293.78 N

10. Calculation of working factor of safety:

$$\text{Working Factor of Safety} = \frac{\text{For the Select Rope}}{\text{Effective Load During Acceleration}}$$

$$= \frac{340000}{34269} = 9.9$$

Working factor of safety neglecting bending load $= \dfrac{340000}{11539.73} = 29$

1.4   Design of Transmission Chains and Sprockets

The normal perform of a chain sprocket is not only to drive or be driven by the chain, but also to guide and support it in its intended path. Sprockets factory mode from good quality iron castings are suitable for the majority of applications. For arduous duty it may be necessary to use steel sprockets having a 0.4 % carbon content. For extremely arduous duty the tooth flanks should be flame hardened. There are other materials which may be specified for particular requirements. Stainless steel as an example is used in high temperature or corrosive conditions.

Sprockets are usually of three main types:

* One piece sprockets of steel or cast iron.

* Two pieces split sprockets.

* Sprockets with bolt-on-tooth segments.

The vast majorities of sprockets in use are of the one piece cast iron or fabricated steel design and are usually parallel or taper keyed to a through shaft. In this case it is necessary to remove the complete shaft to be able to remove the sprockets. If the sprockets and shaft have been in place for a number of years or the shaft is in hostile conditions, it may be more economical to replace the complete shaft assembly, rather than try to remove the sprockets from the existing shaft.

If fast detachability is critical without dismantling shafts or bearings the sprockets may be of the split type. These are made in 2.05 sections and the mating faces machined to allow accurate assembly with the shaft in place. After removal of the chain, the sprocket can be dismantled and a new one assembled around the shaft. This type of sprocket is particularly useful on multi-strand conveyors where long through shafts are used. Considerable expense can be saved in sprocket replacement time.

Sprockets with removable tooth segments are particularly useful where sprocket tooth wear is much more rapid than chain wear. With this kinds of sprocket, segments of teeth can be replaced one at a time without having to disconnect or remove the chain from the sprockets, thus considerable expense and downtime can be saved.

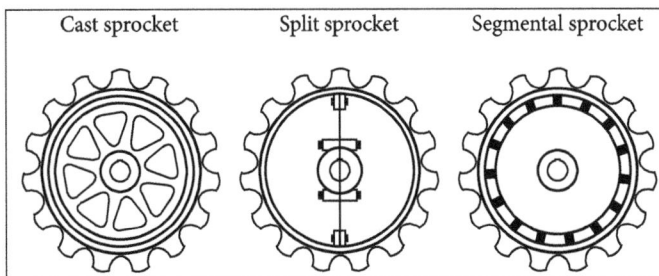

A shafts, whether or not they area unit through shafts or of the stub type, should be of such proportions and strength that sprocket alignment remains unimpaired below load. Shaft sizes should be selected taking under consideration combined bending and torsional moments.

Parts of Roller Chain

- Special profile plates.
- Roller.
- Pins, etc.
- Bushes.
- Sprocket teeth.

Application of Chain Drive

- Used to transmit motion and power from one shaft to another shaft, where center distance is short as in bicycles, motor cycles, conveyor, etc.
- No slip takes place.
- It gives fewer loads on shaft.
- It transmits more power than belt.

Limitation

- Production cost of chain is high.
- It needs accurate mounting of lubrication and slack adjustment.

Problems

1. Let us Design a chain drive to transmit 6 kw at 900 rpm of a sprocket pinion. Speed reduction is 2:5:1. Driving motor is mounted on an adjustable base. Assume that load is steady, drive is horizontal and service is 16 hours/day.

Given data:

- $N = 6\,kw$
- $N_1 = 900\,rpm$
- $i = 2.5$

Solution:

1. Transmission ratio, $i = 2.5$.

$$N_2 = \frac{N_1}{N_{i1}}$$

$$= \frac{900}{2.5}$$

$$= 360 \text{ rpm}$$

2. To find Z1:

$$Z_1 = 27 \text{ (Chosen)}$$

$$Z_2 = i \times 2$$

$$= 2.5 \times 27$$

$$Z_2 = 68$$

3. Standard Pitch (P):

Since the center distance is not given, we have to assume the initial center distance, say a = 500 mm.

$$P_{max} = \frac{a}{30} = \frac{500}{30} = 16.6 \text{ mm}$$

$$P_{min} = \frac{a}{50} = \frac{500}{50} = 10 \text{ mm}$$

The standard pitch P = 15.875 mm

4. Selection of chain:

Assume, the chain to be simplex. The 10 A – 1/250 chain number is chosen.

5. Calculation of total load on the driving side (PT):

$$PT = P_t + P_c + P_s$$

$$P_t = \frac{1020N}{X} = \frac{1020 \times 6}{V}$$

$$V = \frac{2 \times P \times N_1}{60 \times 1000} = \frac{27 \times 15.875 \times 900}{60 \times 1000}$$

$$= 6.42 \, \text{m/s}$$

$$P_t = \frac{1020 \times 6}{6.42} = 953.27 \text{ N}$$

$P_t = 955.27$ N

$P_c = mV^2$

m = 1.01 kg/m

$P_c = 1.01 \times (6.42)2$

    = 41.6 N

$P_s = k.w.a$

k = 6 (for Horizontal)

$\omega = mz$

    = 1.01 × 9.81 = 9.908 N/m

9 = 0.5 m

$P_s = 6 \times 9.908 \times 0.5$

    = 29.72 N

Total load:

PT = 953.27 + 41.6 + 29.72

PT = 1024.59 N

6. Service factor:

$k_s = k_1 \times k_2 \times k_3 \times k_4 \times k_5 \times k_6$

$k_s = 1 \times 1 \times 1 \times 1.25 \times 0.8 \times 1.25$

    = 1.25

7. Design load = PT × k_s

        = 1024.59 × 1.25

        = 1280.73 N

8. Working factor of safety = $\dfrac{\text{Breaking Load}}{\text{Design Load}}$

$$= \frac{222.00}{1280.73}$$

$$= 17.3$$

9. Bearing stress in the roller:

$$\sigma = \frac{P_t \times k_s}{A}$$

$$= \frac{953.17 \times 1.25}{70}$$

$$= 17.02 \text{ N} / \text{mm}^2$$

10. Actual length of chain (α):

$$l_{10} = 2a_P + \left(\frac{Z_1 + Z_2}{L}\right) + \frac{\left[(Z_2 - Z_1)/2\pi\right]^2}{a_P}.$$

$$a_P = \frac{500}{15.875} = 31.496.$$

$$l_P = 2 \times (31.496) + \left(\frac{27 + 68}{2}\right) + \left[\frac{(68 - 27)/2\pi^2}{31.496}\right]$$

$$= 110.49 + 1.35$$

$$= 111.8$$

$$= 112 \text{ C Round off to an even number}$$

$$\alpha = l_P \times p$$

$$= 112 \times 15,875$$

$$= 1778 \text{ mm}$$

11. Exact center distance:

$$a = \frac{e + \sqrt{e^2 - 8m}}{4} \times 4$$

$$a = l_P - \left(\frac{Z_1 + Z_2}{2}\right)$$

$$= 112 - \left(\frac{27 + 68}{2}\right)$$

$$= 64.5$$

$$m = \left[\frac{Z_2 - Z_1}{2\pi}\right]^2 = \left[\frac{68 - 27}{2\pi}\right]^2 = 45.58.$$

$$a = \frac{64.5 + \sqrt{64.5^2 - (8 \times 42.583)}}{4} \times 15.875$$

$$= 501.2 \text{ mm}$$

$$A \times a = 0.01 \times a$$

$$= 0.01 \times 501.2$$

$$= 5.01$$

∴ Exact center distance $= 501.2 - 5.01$

$$= 496.19 \text{ mm}$$

10. Sprocket diameter:

$$p_{cd} = \frac{P}{\sin(180/21)} = \frac{15.875}{\sin(180/27)}$$

$$= 136.74 \text{ mm}$$

$$d_a = d_1 + 0.8 \, d_r$$

$$d_r = 10.16 \text{ mm}$$

$$d_a = 136.74 + 10.8 \times 10.163 = 144.87 \text{ mm}$$

For larger sprocket:

$$P_{CD} = \frac{P}{\sin(180/Z_2)} = \frac{15.875}{\sin(180/68)}$$

$$= 343.73 \text{ mm}$$

$$d_a = d_2 + 0.8 \, d_r$$

$$= 343.73 + 0.8 \times 10.16$$

da=351.85 mm

2. A truck equipped with a 9.5 kw engine uses a roller chain as the final drive to the rear axle. The driving sprocket runs at 900 rpm and (the driven sprocket at 400 rpm with a center distance of approximately 600 mm. Select the roller chain.

Given data:

- Power = 9.5 kw
- Driving speed = 900 rpm
- Driven speed = 400 rpm
- Center distance = 600 mm

Solution

- Select a roller chain.

- Select the number of teeth on the pinion sprocket.

- Determine the (Z_2) and check weather $hq(Z_2)$ is less than max·

- Determine the large of chain pitch using the formula.

- Assume initially simplex (or) duplex chain.

- Evaluate the total Load, $(\varphi_T) = P_t + P_c + P_s$.

- Evaluate service factor.

- Calculate Design Load as, $P_T \times k_s$.

- Check the bearing stress in the roller.

- Calculate the length of chain.

- Find out p_{cd} of sprocket.

1. Determination of the transmission ratio (i):

$$i = \frac{\text{Driver Speed}}{\text{Driven Speed}} = \frac{900}{400} = 2.25$$

2. Selection of teeth on the driver sprocket (Z_1):

Select Z_1 = No. of teeth in driver 25 (for i = 2 to 3).

3. Number of teeth on the driven sprocket (Z_2):

$Z_2 = iZ_1 = 2.25 \times 25$

$\quad = 56.25\ \Omega\ 60$ numbers.

Z_2 max = 100 to 120

So, Z_2 = 60 is recommended.

4. Optimum center distance:

a = (30 to 50) Pitch.

a = Center distance = 600 mm is given.

$$600 = 30\,\text{pitch} = P_{max} = \frac{600}{300} = 20 \text{ mm}$$

$$600 = 500\,\text{pitch} = P_{min} = \frac{600}{50} = 12 \text{ mm}$$

Selecting, higher pitch value of 15.875 mm (standard).

5. Assume the chain to be duplex, the chosen no is 10 A - 2/DR 50.

6. Total load on the driving side of the chain (Pfot):

$$P_{total} = P_{tang} + P_{cent} + P_{seq}$$

$$V = \frac{\text{Number of Teeth on the Sprocket} \times \text{Pitch} \times \text{rpm}}{60 \times 1000}$$

$$= \frac{25 \times 15.875 \times 900}{60 \times 1000}$$

$$= 5.95 \text{ m/s}$$

$$P_{tan\,gential} = \frac{1020 \cdot N}{V}$$

$$= 1020 \times \frac{9.5}{5.95}$$

Where, N = Power = 9.5 kW

$$P_{tang} = 1630 \text{ N}$$

$$P_C = \text{Centrifuges Tension} = \frac{\omega \cdot v^2}{g} = mv^2$$

m = 1.78 kg/m

$$S_t P_c = 1.78 \times (5.95)^2$$

$$P_c = 63 \cdot N$$

Tension due to sagging:

$$P_S = k.w.a$$

k=6 – for horizontal drive

w = mg = 17.8 N/m

a = 600 mm = 0.6 m. (given)

$$P_S = 6 \times 17.8 \times 0.6$$

$$P_s = 64.08 \text{ N}$$

$$P_{total} = P_{tan} + P_c + P_{sage} = 1630 + 63 + 64$$

$$P_{total} = 1757 \text{ N}$$

7. Design Load = $k_s \times$ ptotal.

$k_s = k_1 \cdot k_2 \cdot k_3 \cdot k_4 \cdot k_5 \cdot k_6.$

$k_1 = 1.25$ (For variable load with mild short).

$k_2 = 1$ (For adjustable support).

$k_3 = 1$ (For a = (30 to 50/P).

$k_4 = 1$ (Drive is Horizontal).

$k_5 = 1$ (For Drop Lubrication).

$k_6 = 1.25$ (For 16 hr/day working).

$k_s = 1.25 \times 1 \times 1 \times 1 \times 1.25$

= 1.5625

kS Ω 1.6

Design Load = $k_s \times 1757$

= 1.6 × 1757

Design Load = 2811.2 N

Factor of safety = Break Load/Design Load $= \dfrac{44400}{2811.2} = 15.7839$

From the table, actual factor of safety (15.7839) i) More than the available factor of safety (11) in the table.

8. Bearing Stress on Roller:

$$\text{Induced Stress} = \sigma = \frac{P_L \times k_s}{A}$$

Where, from table = A = 140 mm²

Where Area (A) = 140 mm² for the selected type chain $= \dfrac{2811 \times 1.6}{140}$

σinduced = 32 N/mm²

9. Length Chain:

$$L_p = 2a_p + \left(\frac{Z_1 + Z_2}{2}\right) + \left(\frac{Z_2 - Z_1}{2\pi}\right)/a_p.$$

Where,

$$a_p = \frac{\bar{a_o}}{P} = \frac{\text{Center Distance}}{\text{Pitch}} = \frac{600}{15875} = 37.79 \text{ mm.}$$

$$L_P = 2(37.79) + \left(\frac{25+60}{2}\right) + \left(\frac{60-25}{2 \times 3.14}\right)^2 / 37.79$$

$$= 75.59 + 42.5 + 0.821$$

$$= 118.9$$

$L_p \ \Omega \ 119 \text{ Links}$

$L_p = 119 \text{ links}$

Actual length of chain = 119 × pitch

$$= 119 \times 15.875$$

$L_p = 1889 \text{ mm}$

10. Sprocket diameter:

$$P_{cd} \text{ of Smaller Sprocket} (d_1) = \frac{P}{\sin(180/25)}$$

$$= \frac{15.875}{\sin(180/25)}.$$

$P_{cd1} = 1.26 \text{ mm}$

$$P_{cd} \text{ of Larger Sprocket} (d_2) = \frac{P}{\sin(180/22)}$$

$$= \frac{15.875}{\sin(180/60)} = 303 \text{ mm}$$

Result:

- Chain type 10A - 2/DR50.
- Length of chain = 1889 mm.
- Pcd_1 = 126 mm.
- Pcd_2 = 303 mm.

- Design Load = 2811 N.

- Z_1 = 25 nos.

- Z_2 = 60 nos.

3. Design a chain drive to drive a centrifugal compressor from an electric motor 15 kW at 1000 rpm. The speed reduction ratio required is 2.5. The compressor to work for 16 hours a day. State solutions for common problems encountered in continuous operation of the drive.

Given:

Application, centrifugal compressor:

- p = 15 kW.

- N_1 = 1000 rpm.

- P = 2.5.

- Service = 16 hours/day.

Solution

Step 1:

$$\text{For i} = 2.5, Z_1 = 26$$

$$Z_2 = 65$$

Step 2:

$$\text{For } N_1 = 1000 \text{ rpm, P} = 15.875 \text{ mm}$$

$$d_1 = \frac{P}{\sin\dfrac{180}{21}} = 131.7 \text{ mm.}$$

$$d_2 = \frac{P}{\sin\dfrac{180}{22}} = 328.58 \text{ mm}$$

$$\therefore a = 40 \, P = 635 \text{ mm}$$

Step 3:

$$a = 635 \text{ mm}$$

Step 4:

Breaking load:

$$Q = \frac{N \times 100 \times n \, k_s}{V}$$

$$V = \frac{\pi d_1 N_1}{60} = 6.89 \text{ m/s}$$

$n = 11$

$k_s = 1.25$ (16 hrs/day)

$\therefore Q = 2993.47$ kgf

Qstd = 4540 kgf

Step 5:

Chain No: Rolex DR 1595

$p = 15.875$ mm

$D_r = 10.16$ mm

$W = 9.85$ mm

$D_p = 5.08$ mm

$G = 14.3$ mm

$p_t = 16.59$ mm

$A_1, A_2 = 30.9$ mm

Bearing area = 1.34 m²

Weight/meter = 1.82 kgf/m

Step 6:

Induced stress:

$$\sigma = \frac{N \times 100 \, k_s}{AV}$$

$\sigma = 203.08$ kgf/cm²

σstd = 224 kgf/cm²

$\sigma < \sigma_{std}$

Step 7:

$$[n] = \frac{\theta}{\sum P}$$

$$= \frac{\theta}{P_t + P_c + P_s}$$

$$P_t = \frac{100\,N}{V} = 217.7\,\text{kgf}$$

$$P_c = \frac{WV^2}{g} = 8.8\,\text{kgf}$$

$$p_s = k.w.a = 6.93\,\text{kgf}$$

$$[n] = 19.45$$

Step 8:

$$l_p = 2a_p + \frac{Z_1 + Z_2}{2} + \frac{\left(\frac{Z_2 - Z_1}{2\pi}\right)^2}{a_p}$$

$$l_p = 126$$

∴ Length of the chain, $l = l_p \cdot P$

$$= 2000 \text{ mm}$$

4. Let us design a chain drive to actuate a compressor from 15 kW electric motor running at 1,000 rpm, the compressor speed being 350 rpm. The minimum center distance is 500 mm. The compressor operates 15 hours per day. The chain tension may be adjusted by shifting the motor.

Given: Compressor, p = 15 KW, n_1 = 1000 rpm, n_2 = 350 rpm, a = 500 mm, 15 hours/day.

Solution

$$i = \frac{n_1}{n_2} = 2.85 \approx 3$$

For i = 3, z_1 = 25, z_2 = 75

Diameters of sprockets:

$$d_1 = \frac{P}{\sin\dfrac{180}{z_1}} = 101.3 \text{ mm}$$

$$d_2 = \frac{P}{\sin\dfrac{180}{z_2}} = 303.2 \text{ mm}$$

Since, $a_{opt} = 40 \Rightarrow p = \dfrac{500}{40} = 12.5 \approx 12.7 \text{ mm}$

Breaking load:

$$N = \frac{Q.V}{100\,n\,k_s}$$

$$k_s = 1 \times 1 \times 1 \times 1 \times 1 \times 1.25 = 1.25$$

$$V = \frac{\pi \times 101.3 \times 10^{-3} \times 1000}{60} = 5.3 \text{ m/s}$$

$$\therefore 15 = \frac{Q \times 5.3}{100 \times 11 \times 1.25} \quad [\text{For} \, 1000 \text{ rpm, n} = 11]$$

Q = 3891.5 kgf

Q_{std} = 4540 kgf

Roller Chain No.: TR1278

P = 12.7 mm

D_r = 8.51mm

W = 8 mm

D_p = 4.45 mm

G = 11.7 mm

A = 1.5 cm²

Weight/m = 1.95 kgf

Bearing stress:

$$N = \frac{\sigma A V}{100 K_S}$$

$$15 = \frac{\sigma \times 1.5 \times 5.3}{100 \times 1.25}$$

$$\sigma = 235.84 \ kgf / cm^2$$

Load on shaft:

$$Q_o = 1.15 \times \frac{100 \times 15}{5.3}$$

$$= 325.47 \ kgf$$

5. The reduction of speed from 360 rpm to 120 rpm is desired by the use of chain drive. The driving sprocket has 10 teeth. Let us calculate the number of teeth on the driven sprocket. If the pitch radius of the driven sprocket is 250mm and the center to center distance between the two sprockets is 400mm, and determine the pitch and length of the chain.

Given Data:

$$N_1 = 120 \ rpm$$

$$N_2 = 300 \ rpm$$

$$Z_1 = 10$$

$$a = 400 \ mm$$

Solution

$$i = \frac{N_1}{N_2} = \frac{120}{360} = \frac{360}{120} = 3$$

$$i = \frac{Z_2}{Z_1}$$

$$i = 3$$

$$Z_2 = i \times Z_1$$

$$= 3 \times 10 = 30$$

Section of pitch (P):

$$a = (30 - 50)P$$

$$\text{Maximum pitch, } P_{max} = \frac{a}{50} = \frac{400}{30} = 13.3$$

$$\text{Minimum pitch, } P_{min} = \frac{a}{50} = \frac{400}{50} = 8$$

Any standard pitch, between 8 mm and 133 mm can be chosen.

Standard pitch = 9.525 mm.

Length of the chain (L):

$$L = l_p \times P$$

$$l_p = 2\left(\frac{a}{P}\right) + \left(\frac{Z_1 + Z_2}{2}\right) + \left(\frac{Z_2 - Z_1}{2\pi}\right)^2 \times \left(\frac{P}{a}\right)$$

$$= 2\left(\frac{400}{9.525}\right) + \left(\frac{10+30}{2}\right) + \left(\frac{30-10}{2\pi}\right)^2 \times \left(\frac{9.525}{400}\right)$$

$$= 103.98 + (3.18 \times 0.02)$$

$$= 103.98 + 0.06$$

$$l_p = 104.04 \text{ mm}$$

$$L = 104 \times 9.525$$

$$L = 990.6 \text{ mm.}$$

Spur Gears and Parallel Axis Helical Gears

2.1   Speed Ratios and Number of Teeth

Module

It is the ratio of pitch circle diameter to the number of teeth; it is denoted by 'm'.

$$m = \frac{D}{T}$$

Law of Gearing

The common normal at tooth profile at point of contact should always pass through fixed point called pitch point in order to obtain a constant velocity ratio.

Gear Tooth Subjected to Dynamic Load

In a gear tooth, dynamic loads are due to the following reasons:

- Inaccuracies of tooth spacing.
- Elasticity of parts.
- Deflection of teeth under load.
- Irregularities in tooth profiles.
- Misalignment between bearings.
- Dynamic unbalance of rotating parts.

2.2   Force Analysis

The Spur Gear's transmission force F_n, which is normal to the tooth surface as in figure can be resolved into a tangential component, F_t, and a radial component, F_r. There will be no axial force, F_x.

The directions of the forces acting on the gears are shown in given:

$$F_t = F_n \cos \alpha'$$
$$F_t = F_n \sin \alpha'.$$

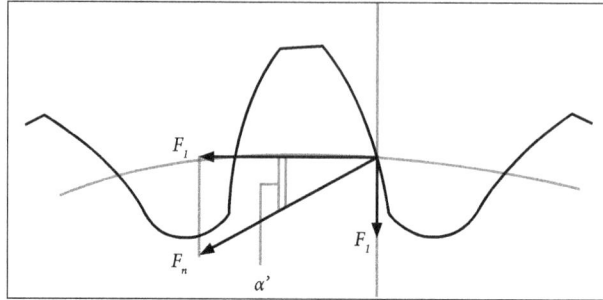

Forces Acting on a Spur Gear Mesh.

Problems

The pitch circles of a train of spur gears are shown in figure. Gear Q receives 3.5 kW power at 700 rpm through its shaft and rotates in clockwise direction. Gear B is the idler gear while gear C is the driven gear. The number of teeth on gears A, B and C are 30, 60 and 40 respectively, while the module is 5 mm. Calculate the torque on each gear shaft; and the components of gear tooth forces. Draw a free-body diagram of forces and determine the reaction on the idler gear shaft. Assume the 20° in-volute system for the gears.

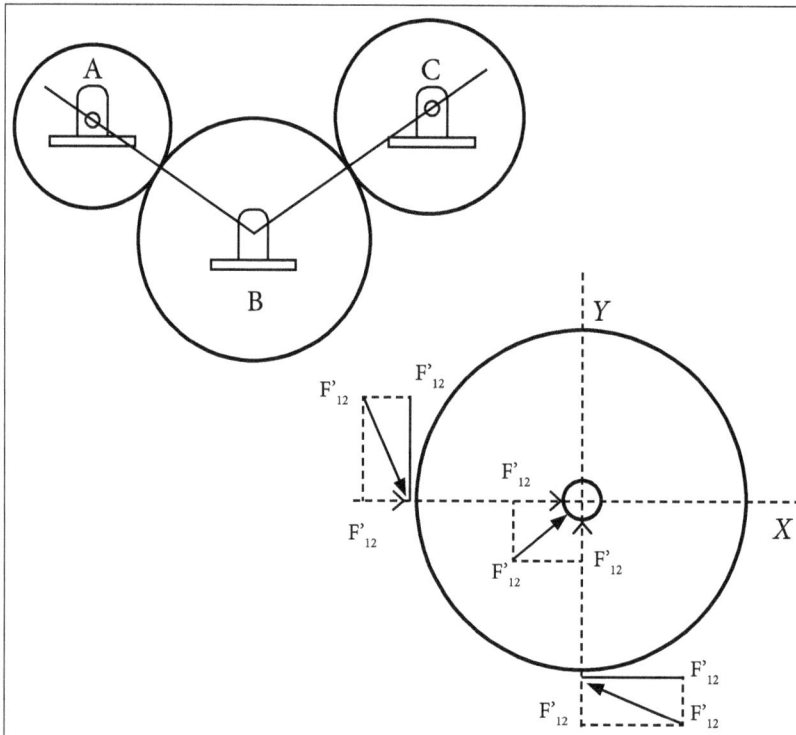

Given Data:

- N_1 = 700 rpm
- P = 3.5 kW
- φ = 20°
- m = 5 mm
- Z1 = 30; Z2 = 60 and Z3 = 40

Solution

d_1= mz_1 = 5 × 30 ⇒ 150 mm

d_2 = mz_2 = 5 × 60 ⇒ 300 mm

d_3 = mz_3 = 5 × 40 ⇒ 200 mm

$$V_1 = \frac{\pi d_1 N_1}{60} = \frac{\pi \times \left(150 \times 10^{-3}\right) \times 70\varphi}{6\varphi}.$$

All the above 3 gears have same pitch line velocity (V).

$$V_1 = 5.49 \text{ m/s}$$

Considering Piston 1 and Gear 2:

Let the tangential force of piston 1 on gear 2 is F_{12}^t and given by:

$$P = F_t \times V$$

$$F_{12}^t = \frac{P}{V} = 3.5 \times 10^3 / 5.49 = 637.52 \text{ N}.$$

∴ Radial force of piston 1 on gear 2:

$$F_{12}^t = F_{12}^t \times \tan\varphi$$
$$= 637.52 \times \tan 20°$$
$$= 232.03 \text{ N}$$

Then, the resultant force of piston 1 on gear 2:

$$F_{12} = \frac{F_{12}^t}{\cos\varphi} = \frac{637.52}{\cos\varphi}$$
$$= 678.43 \text{ N}$$

The torque transmitted by piston 1:

$$T_1 = \frac{60 \times P}{2\pi N_1}$$

$$= \frac{60 \times 3.5 \times 10^3}{2 \times \pi \times 700}$$

$$= 47.74 \text{ N} - \text{m}$$

Considering Gear 2 and 3:

Since, gear 2 is an idler, it transmits no torque (power) to its shaft.

$\therefore$ Torque transmitted by gear $J_2 = 0$.

Since the gear 2 is an idler. Whatever torque it receives from pinion 1 is transmitted to gear 3. Therefore, the tangential component between gear 2 and 3 must be equal to the tangential component between gears 1, and 2.

$$F^t_{32} = F^5_{12} \Rightarrow 237.52 \text{ N}$$

$$F^r_{32} = F^r_{12} \Rightarrow 232.03 \text{ N}$$

$$F_{32} = F^r_{12} \Rightarrow 678.43 \text{ N}$$

$\therefore$ The same power is transmitted from piston 1 to gear 3.

$$\frac{2\pi N_1 T_1}{60} = \frac{2\pi N_3 T_3}{60}.$$

$$N_1 T_1 = N_3 T_3.$$

Torque transmitted by gear 3:

$$T_3 = T_1\left(\frac{N_1}{N_3}\right) = T_1\left(\frac{Z_3}{Z_1}\right)$$

$$T_3 = 47.74\left(\frac{4\varphi}{3\varphi}\right) \left[\because i = \frac{N_1}{N_3} = \frac{Z_3}{Z_1}\right]$$

$$T_3 = 63.65 \text{N} - \text{m}$$

Reaction on shaft ideal gear:

The shaft reactions in the x and y direction:

$$R^x_{b2} = -\left(F^t_{12} + F^r_{32}\right).$$

$$R^y_{b2} = -\left(F^r_{12} + F^t_{32}\right).$$

Hence, (−) negative sign is for reaction. Because the reaction is always opposite in direction to applied force.

$$R_{b2}^x = -(637.52 + 232.03) = 869.5 \quad ...\left[U\sin g\, Sign\, Convection\,(+) \rightarrow; (-)*\right].$$

$$R_{b2}^y = -(232.03 + 637.52) = 869.5 \quad ...\left[U\sin g\, Sign\, Convection\,(+) \rightarrow; (-)*\right].$$

(i.e.,) The resultant shaft reaction is given by:

$$R_{b2} = \sqrt{\left(R_{b2}^x\right)^2 + \left(R_{b2}^y\right)^2}$$
$$= \sqrt{(869.5)^2 + (869.5)^2}$$
$$= 1739\ N$$

2.2.1 Tooth Stresses

Gears are used for a wide range of industrial applications. They have varied application starting from textile looms to aviation industries. They are the most common means of transmitting power. They change the rate of rotation of machinery shaft and also the axis of rotation. For high speed machinery, such as an automobile transmission, they are the optimal medium for low energy loss and high accuracy. Their function is to convert input provided by prime mover into an output with lower speed and corresponding higher torque.

Toothed gears are used to transmit the power with high velocity ratio. During this phase, they encounter high stress at the point of contact. A pair of teeth in action is generally subjected to two types of cyclic stresses:

- Bending stresses inducing bending fatigue.

- Contact stress causing contact fatigue.

Both these types of stresses may not attain their maximum values at the same point of contact. However, combined action of both of them is the reason of failure of gear tooth leading to fracture at the root of a tooth under bending fatigue and surface failure, due to contact fatigue. When loads are applied to the bodies, their surfaces deform elastically near the point of contact. Stresses developed by Normal force in a photo-elastic model of gear tooth. The highest stresses exist at regions where the lines are bunched closest together.

The highest stress occurs at two locations:

- At contact point where the force Facts.

- At the fillet region near the base of the tooth.

2.2.2 Dynamic Effects

Reducing the dynamic loading and noise of gear systems has been an important concern in gear design. Generated from gearing is basically due to gearbox vibration excited by the dynamic load. This vibration is transmitted through shafts and bearings to noise-radiating surfaces on the gearbox exterior. Minimizing gear dynamic loading will reduce gear noise. Many researchers have found that the noise the principal source of gear system vibration is the unsteady component of the relative angular motion of meshing gear pairs. The static transmission error describes this displacement type of vibratory excitation. The variation of gear-pair meshing tooth stiffness, which causes static transmission error, teeth error and wear, is, primarily due to the periodic alternation in the numbers of contacting teeth. Secondary effects include tooth profile modification, machining and wear.

2.2.3 Fatigue Strength

The majority of engineering failures are caused by fatigue. Fatigue failure is defined as the tendency of a material to fracture by means of progressive brittle cracking under repeated alternating or cyclic stresses of intensity considerably below the normal strength. Although the fracture is of a brittle type, it may take some time to propagate, depending on both the intensity and frequency of the stress cycles. Nevertheless, there is very little, if any, warning before failure if the crack is not noticed. The number of cycles required to cause fatigue failure at a particular peak stress is generally quite large, but it decreases as the stress is increased. For some mild steels, cyclical stresses can be continued indefinitely provided the peak stress (sometimes called fatigue strength) is below the endurance limit value.

A good example of fatigue failure is breaking a thin steel rod or wire with your hands after bending it back and forth several times in the same place. Another example is an unbalanced pump impeller resulting in vibrations that can cause fatigue failure.

The type of fatigue of most concern in circuit cards, gasoline, diesel, gas turbine engines and many industrial applications is thermal fatigue. Thermal fatigue can arise from thermal stresses produced by cyclic changes in temperature.

Fundamental requirements during design and manufacturing for avoiding fatigue failure are different for different cases and should be considered during the design phase.

2.3   Factor of Safety

It is defined as the ratio of the ultimate strength of a member or piece of material to the actual working stress or the maximum permissible stress when in use.

The factor of safety is often specified in a design code or standard, such as:

- American Institute of Steel Construction (AISC) – steel buildings & bridges.

- American Society of Mechanical Engineers (ASME) – pressure vessels, boilers, shafts.

- American Concrete Institute (ACI).

- National Forest Products Association (NFPA) – wood structures.

- Aluminum Association (AA) – aluminum buildings & bridges.

- Codes often specify a minimum factor of safety.

- Designer's responsibility to determine if a code or standard applies. Codes are often specified by law (BOCA, UBC, etc.).

Factors which Affect the Factor of Safety

Material Strength Basis:

- Brittle Materials – use ultimate strength.

- Ductile Materials – use yield strength.

Manner or Loading:

- Static – applied slowly; remains applied or is infrequently removed.

- Repeated – fatigue failure may occur at stresses lower than static load failure.

- Impact – high initial stresses develop.

- Possible misuse – designer must consider any reasonable foreseeable use & misuse of product.

- Complexity of stress analysis – the actual stress in a part isn't always known.

- Environment – temperature, weather, radiation, chemical, etc.

2.4   Gear Materials

Materials used for Manufacturing of Gears

The material used for the manufacture of gears depends upon the strength and service conditions like wear, noise etc. The gears may be manufactured from metallic or non-metallic materials. The metallic gears with cut teeth are commercially obtainable in cast iron, steel and bronze. The non-metallic materials like wood, rawhide, compressed paper and synthetic resins like nylon are used for gears, especially for reducing noise.

The cast iron is widely used for the manufacture of gears due to its good wearing properties, excellent machinability and ease of producing complicated shapes by casting method. The cast iron gears with cut teeth may be employed, where smooth action is not important.

The steel is used for high strength gears and steel may be plain carbon steel or alloy steel. The steel gears are usually heat treated in order to combine properly the toughness and tooth hardness.

Materials	Condition	Brinell Hardness Number	Minimum Tensile Strength (kglcm2)
Malleable cast iron:			
a) White heart castings, Grade	---	217 max.	280
b) Black heart castings, Grade B	---	149 max.	320
Cast Iron:			
a) Grade 20	As cast	179 min.	200
b) Grade 25	As cast	197 min.	250
c) Grade 35	As cast	207 min.	250
d) Grade 35	Heat treated	300 min.	350
Cast steel:	---	145	550
Carbon steel:			
a) 0.3% carbon	Normalised	143	500
b) 0.3% carbon	Hardened and tempered	152	600
c) 0.4% carbon	Normalised	152	580
d) 0.4% carbon	Hardened and tempered	179	600
e) 0.35% carbon	Normalised	201	720
f) 0.55% carbon	Hardened and tempered	223	700
Carbon chromium steel:			
a) 0.4% carbon	Hardened and tempered	229	800
b) 0.55% carbon	Hardened and tempered	225	900
Carbon manganese steel:			
a) 0.27% carbon	Hardened and tempered	170	600
b) 0.37% carbon	Hardened and tempered	201	700

Manganese molybdenum steel:			
a) 35 M_n2 M_o28	Hardened and tempered	201	700
b) 35 M_n2 M_o45	Hardened and tempered	229	800
Chromium molybdenum steel:			
a) 40 C_r1 M_o28	Hardened and tempered	201	700
b) 40 C_r1 M_o60	Hardened and tempered	248	900

- Metallic Materials:
  - Cast iron.
  - Steel.
  - Bronze.
- Non-Metallic Material:
  - Wood.
  - Rawhide.
  - Synthetic resins.
  - Compressed papers.

2.5 Design of Straight Tooth Spur and Helical Gears Based on Strength and Wear Considerations

In the design of a gear drive, the following data is usually given:

- The power to be transmitted.
- The speed of the driving gear.
- The speed of the driven gear or the velocity ratio.
- The centre distance.

The following requirements must be met in the design of a gear drive:

- The gear teeth should have sufficient strength so that they will not fail under static loading or dynamic loading during normal running conditions.
- The gear teeth should have wear characteristics so that their life is satisfactory.

- The use of space and material should be economical.

- The alignment of the gears and deflections of the shafts must be considered because they effect on the performance of the gears.

- The lubrication of the gears must be satisfactory.

Problems

A 27.5 kW power is transmitted at 450 rpm to a shaft running at approximately 112 rpm through a spur gear drive. The load is steady and continuous. Design the gear drive and check the design. Assume the following materials: Pinion-heat treated cast steel; Gear-High grade cast iron.

Data:

- $P = 27.5$ kW

- $N_1 = 450$ rpm

- $N_2 = 112$ rpm

- $\Phi = 20°$

- $a = 375$ mm (Assume)

Find: Design the spur gear drive.

Solution

Since, the materials of pinion and gear are different, first we have to know $[\sigma b_1] \cdot y1$ and $[\sigma b_2] \cdot y2$ to:

$$\text{Gear Ratio, } i = \frac{N_1}{N_2} = \frac{450}{112} = 4.01.$$

Find out the weaker elements.

Where, i=4.

Assume, $Z_1 = 18$

$$Z_2 = i \times Z_1 = 4 \times 18 = 72$$

Pinion:

Form factor:

$$Y_1 = 0.270$$

$$Z_1 = 18$$

Pinion:

Heat treated cast steel:

$[\sigma_b] = 196$ N/mm² [Static permissible stress].

$$[\sigma_{b1}]y_1 = 196 \times \frac{0.270}{\pi} = 16.85$$

For gear:

$\quad$ $Y_2 = 0.360$

$\quad$ $Z_2 = 72$

Permissible static stress:

$\quad$ $[\sigma_{b2}] \cdot y_2 = 56$ N/mm² for cast iron

$$[\sigma_{b2}]y_2 = 56 \times \frac{0.360}{\pi} = 6.42$$

We find: $[\sigma_{b2}]y_2 < [\sigma_{b1}]y_1$ (i.e).

The gear is weaker than the pinion. Therefore, we have design the gear only.

1. Material selection:

Pinion: heat treated cast steel.

Gear: cast iron.

2. Calculation of module (m):

Since the center distance (a) is gives, we need to equate Fs & Fd to find the Module. Here the Module can be calculated using the relation:

$$a = \frac{m(Z_1 + Z_2)}{2}$$

$$375 = \frac{m(18 + 72)}{2}$$

$\quad$ Module, m = 8.333

The nearest higher standard module under choice from its 10 mm.

3. Calculation of b, d and v:

Face width (b):

$\quad$ b = 10×m

$$= 10 \times 9$$

$$b = 90 \text{ mm}$$

Pitch circle diameter of pinion (d_1):

$$d_1 = m \cdot Z_1$$

$$= 9 \times 18$$

$$d_1 = 162 \text{ mm}$$

Pitch circle diameter of gear (d_2):

$$d_2 = m \cdot Z_2$$

$$= 9 \times 72$$

$$d_2 = 648 \text{ mm} \Rightarrow \text{where } d_2 = 648 \text{ mm}$$

Pitch line velocity (v):

$$\gamma = \frac{\pi d_2 N_2}{60}$$

$$= \frac{\pi \times 648 \times 10^{-3} \times 112}{60}$$

$$= 3.79 \text{ m/s}$$

4. Calculation of beam strength (F_s):

Beam Strength:

$$F_s = \pi \cdot m \cdot b \cdot [\sigma b_2] \cdot y_2$$

$$= 3.14 \times 9 \times 90 \times 56 \times \left(\frac{0.36}{3.14}\right)$$

$$= 16329.6 \text{ N}$$

5. Calculation of Dynamic Load (F_d):

Dynamic load:

$$F_d = F_t + \frac{21 V (bc + F_t)}{21 V + \sqrt{bc + F_t}}$$

$$F_t = \frac{D}{V}$$

$$= \frac{27.5 \times 10^3}{3.79}$$

$$F_t = 7255.9 \text{ N}$$

C = Deformation factor

$\quad$ = 8150 e, for 20° FD

e = 0.20 for Module up to 9 and precision gears.

C = 8150 × e

$\quad$ = 8150 × 0.020

C = 163 N/mm

Then,

$$F_d = 7255.4 + \frac{21 \times 3.79 \times 10^3 \left(9 \times 163 + 7255.9\right)}{21 \times 3.79 \times 10^3 + \sqrt{9 \times 163 + 7255.9}}$$

$$= \frac{694255.611}{79683.396} = 8712.675 \text{ N}$$

6. Check for beam strength (or tooth breakage):

We find $f_d < F_s$.

It means, the gear tooth has adequate beam strength and it will not fail breakage.

Thus the design is satisfactory.

7. Calculation of wear load (f_w):

Maximum wear load:

$$F_W = d_1 \times b \times Q \times K_W$$

Ratio factor:

$$Q = \frac{2i}{i+1}$$

$$= \frac{2 \times 4}{4+1} = 1.6$$

KW = load stress factor.

$\quad$ = 2 N/mm2, for steel (250 BHN), Cast iron and 20° FD.

$F_W = 162 \times 90 \times 1.6 \times 2.$

$F_W = 4665$ Bn.

8. Check for wear:

We find $F_W > F_d$

It means, the gear tooth has adequate wear capacity and it will not wear out

Therefore the design is satisfactory

9. Result:

Basic dimension of pinion and gear:

Module(m) = 9 mm

No. of Teeth:

$Z_1 = 18$ and $Z_2 = 72$

Pitch Circle Diameter:

$d_1 = 162$ mm and $d_2 = 648$ mm

Center Distance:

a = 375 mm

Face Width:

B=90 mm

Height Factor:

$F_o = 2$, for 20° FD

Bottom Clearance:

C=0.25×9=0.25×9=2.25 mm

Tip Diameter:

$da_1 = (Z_1 + 2F_o)m = (18 + 2 \times 2)9 = 198$ mm

$da_2 = (Z_2 + 2F_o)m = (72 + 2 \times 2)9 = 684$ mm

Root Diameter:

$dF_1 = (Z_1 - 2F_o)m - 2c$

$= (18 - 2 \times 2)9 - 2 \times 2.25$

$$= 1215 \text{ mm} \Rightarrow 121.5 \text{ mm}$$

$$dF_2 = (Z_2 - 2F_o)m - 2C$$

$$= (Z_2 - 2 \times 2)9 - 2 \times 2.25$$

$$= 607.5 \text{ mm}$$

Problem

In a spur gear drive for a stone crusher, the gears are made of C40 steel. The pinion is transmitting 20 kW at 1200 rpm. The gear ratio is 3. Gear is to work 8 hrs. per day, six days a week and for 3 years. Design the drive.

Given data:

Spur gear drive for stone crushes:

- Power = 20 kW

- N = 1200 rpm

- Gear ratio = 3

Working hours = 8 hours/day and 6 days/work for 3 years

$$= (8 \times 6) \, 4 \times 12 \times 3$$

$$= 6912 \text{ hrs.}/3 \text{ years } \Omega \text{ 1000 hrs}$$

Solution

1. Selection of material:

C40 surface hardened to 55RC.

Core hardness 350BHN:

$$\sigma_v = 720 \text{ N/mm}^2; \sigma_y = 360 \text{ N/mm}^2$$

2. Calculation of design stress:

$[\sigma_b]$–Design bending stress

$$\left[\sigma_b\right] = \frac{1.4 \times k_{be}}{n \cdot k_\sigma} \sigma - 1$$

Where,

$$\sigma_{-1} = 0.25 \, (\sigma_v + \sigma_y) + 50$$

$$= 0.25 \, (720 + 360) + 50 = 320 \text{ N/mm}^2$$

$\sigma_{-1} = 320 \text{ N/mm}^2$

Select n = 2.5

$k_a = 1.5$ to get:

k_{be} N=life=10,000×60×1200

$\qquad = 72 \times 10^7$ cycles

$k_{be} = 1$ (core hardness)

< 350 BHN

Substitute all the value in the (σ_b) equation:

$$[\sigma_b] = \frac{1.4 \times k_{b1}}{n k_\sigma} \sigma - 1$$

$$= \frac{1.4 \times 1 \times 320}{2.5 \times 1.5} = 119.5 \text{ N/mm}^2$$

$[\sigma_c] = -$ design crushing stress.

$[\sigma_c] = Cr \cdot HRC \times k_{ce} \, k_{c1} = 0.585$

$\qquad = 23 \times 55 \times 0.525$

$[\sigma_c] = 740 \text{ N/mm}^2$

Centre distance calculation:

$$a \geq (i+1) 3 \sqrt{\left[\frac{0.74}{[\sigma_c]}\right]^2 \times \frac{E \times [M_t]}{i\psi}}$$

Where,

$$M_t = \frac{P \times 60}{2\pi \times 764} = \frac{20 \times 10^3 \times 60}{2 \times \pi \times 764} = 250 \text{ N-m}$$

Assume:

$\psi = 0.3$; $k \cdot kd = 1.3$; $k_o = 1$; $E = 2 \times 10^5 \text{ N/mm}^2$

$i = 3$; $[M_t] = k_o k \cdot kd \cdot M_t = 1 \times 1.3 \times 250 = 3.25 \text{ N-m}$

$$a \geq (3+1) 3 \sqrt{\left[\frac{0.74}{740}\right]^2 \times \frac{2.1 \times 10^5 \times 325 \times 10^3}{3 \times 0.3}}$$

$a \geq 170 \text{ mm}$

Calculation of module:

$$m = \frac{29}{Z_1 + Z_2}$$

$$= \frac{2 \times 170}{20 + 60} = 4.25 \text{ mm}$$

m Ω4 mm.

Assume, $Z_1 = 20$

$$Z_2 = i \, Z_1$$

$$= 3 \times 20$$

$$Z_2 = 60$$

Revise a:

$$a = \frac{m(Z_1 + Z_2)}{2} = \frac{4 \times (20 + 60)}{2} = 160 \text{ mm}$$

$$\psi = \frac{b}{a} = 0.3$$

$$b = 160 \times 0.3 = 48 \text{ mm}$$

Revise $[M_t]$:

$$d_1 = MZ_1 = 4 \times 20 \times 80 \text{ mm}$$

$$\frac{b}{d_1} = \frac{48}{80} = 0.6$$

Therefore, k = 1

Therefore, k_ℓ = 1.3 (for 1S quality 8)

$[Mt]_{revised} = 1 \times 1 \times 1.3 \times 250 = 325$ N–m

$[M_t]_{revised}$, and, old [Mt], same values no need to recalculate the center distance.

Calculation of induced stress:

$$\sigma_b = \frac{(i+1)[M_t]}{a \, m \, b_y}$$

$$= \frac{(3+1)\left[325 \times 10^3\right]}{160 \times 4 \times 48 \times 0.389}$$

Select y = 0.389 for Z_1=20

σ_b= 108 N/mm² which is less than design value.

$\sigma_b < [\sigma_b]$; 109

109 < 119

So, the design is safe in bending

σ_c = Stress induced in crushing:

$$\sigma_{C_{Induced}} = 0.74\left(\frac{i+1}{a}\right)\sqrt{\frac{i+1}{i^b} \times E \times [M_t]}$$

$$= 0.74\sqrt{\left(\frac{3+1}{3 \times 48}\right) \times 2.1 \times 10^5 \times 325 \times 10^3}$$

$$= 730\ N/mm^2 < [\sigma_C] = 740\ N/mm^2$$

The stress induced in crushing is less than the design crushing stress; so the design is safe in crushing. Hence the design is satisfactory.

Result:

- Material = C40.
- Pinion Number of teeth Z_1 = 20 nos.
- Z_2 = 60 nos.
- Speed Ratio i = 3.
- Pinion dia. d_1 = 80 mm.
- Gear dia. d_2 = 240 mm.
- Center distance a = 170 mm.
- Face width b = 48 mm.
- ψ = 0.6.

Problem

Let us design a spur gear drive for a heavy machine tool with moderate shocks. The pinion is transmitting 18 kW at 1200 rpm with a gear ratio of 3.5. Design the drive and check for elastic stresses and plastic deformation. Make a sketch and label important dimensions arrived.

Given:

$p = 18$ kW

$N_1 = 1200$ rpm

$i = 3.5$

Solution

Step 1: $i = 3.5$ (given).

Step 2: Assume C45 for both pinion and wheel,

$\sigma^{-1} = 2700$ kgf/cm²

$HB = 215$

$[\sigma_b] = 1400$ kgf/cm²

$[\sigma_c] = 5000$ kgf/cm²

$\sigma_y = 3600$ kgf/cm²

Step 3:

Center distance:

$$a \ge (i+1)3\sqrt{\left[\frac{0.74}{[\sigma_c]}\right]^2 \times \frac{E \times [M_t]}{i\psi}}$$

$E = 2.15 \times 106$ kgf/cm²

$$[M_t] = 1.3 \times 97420\frac{kW}{n_1}$$
$$= 1899.69 \text{ kgf / cm}$$

Assume:

$\psi = 1$

$$\left(a \ge (i+1)3\sqrt{\left[\frac{0.74}{[\sigma_c]}\right]^2 \times \frac{E \times [M_t]}{i\psi}}\right) \Rightarrow a \ge 13.25$$

$a = 14$ cm

Step 4:

To avoid interference, assume $Z_1 = 20$

$\qquad Z_2 = 70$

Step 5:

Module:

$$m \geq 1.26\sqrt[3]{\frac{[M_t]}{y[\sigma_b]\psi_m Z_1}}$$

For $Z_1 = 20$ and $y = 0.389$

Assume :

$\qquad \psi_m = 10$

$\qquad m \geq 0.3267$

$\qquad m = 0.4$ cm

Step 6:

$$\sigma_C = 0.74\left(\frac{i+1}{a}\right)\sqrt{\frac{i+1}{ib}E[M_t]}$$

$$\sigma_C = 4606.6 \text{kgf}/\text{cm}^2 < [\sigma_C]$$

$$\sigma_b = \frac{i+1}{a\,m\,by}[M_t]$$

$$\sigma_b = 280.3 \text{ kgf}/\text{cm}^2 < [\sigma_b]$$

Plastic deformation:

$$\sigma_{c\,max} = \sigma_c\sqrt{\frac{M_{t\,max}}{M_t}} = \sigma_c = \sqrt{2} = 6514.7 \text{ kgf}/\text{cm}^2$$

$$[\sigma_{cmax}] = 3.1\,\sigma_y = 11160 \text{ kgf}/\text{cm}^2$$

$$\sigma_{cmax} < [\sigma_{cmax}]$$

$$\sigma_{b\,max} = \sigma_b\frac{M_{t\,max}}{M_t} = 560.6 \text{ kgf}/\text{cm}^2$$

$$[\sigma_{bmax}] = 0.86\,\sigma_y = 2880 \text{ kgf}/\text{cm}^2$$

$$\sigma_{bmax} > [\sigma_{bmax}]$$

Step 7:

$$d_1 = mZ_1 = 8 \text{ cm}$$

$$d_2 = mZ_2 = 28 \text{ cm}$$

Problem

A pair of helical gears subjected to moderate shock loading is to transmit 20kW at 1500 rpm of the pinion. The speed reduction ratio is 4 and the helix angle is 20°. The service is continuous and the teeth are 20° full depth in the normal-plane. For the gear life of 10,000 hours, design the gear drive.

Given Data:

- P = 20 kw

- N_1 = 1500 rpm

- i = 4

- θ = 20°

- ϕ = 20° FD

To find:

Design the helical gear pair.

Solution

1. Gear ratio: i = 4

2. Selection of material: for both pinion and gear, alloy steel 40 Ni 2Cr 10 28 can be selected.

3. Gear life:

Gear life = 10,000 hours/day

$$= 10{,}000 \times 1500 \times 60$$

$$= 90 \times 10^7 \text{ cycles}$$

4. Calculation of initial design torque:

$$M_t = \frac{00 \times P}{2\pi N}$$

$$= \frac{00 \times 20 \times 10^3}{2\pi \times 1500}$$

$$[M_t] = M_t \times K \times K_d \times K_o$$
$$= 127.32 \text{ N} - \text{m}$$

Assuming, $k_d = 1.3$, $k_o = 1.25$ for medium scores.

$$M_t = 127.32 \times 1.3 \times 1.25$$

$$M_t = 206 \times 9 \text{ N} - \text{m}$$

5. Calculation of equation $[\sigma_b]$ and (σ_e):

$E_{eq} = 21.5 \times 105 \text{ N/mm2}$ [For pinion and steel].

To find:

$$\sigma_b = \frac{1.4 \times k_{bl}}{n \times k_\sigma} \times \sigma - 1$$

6. Calculation of center distance:

$$a_1(i+1)3\sqrt{\left[\frac{0.7}{[\sigma_1]}\right]^2 \times \frac{E_{eq} \times [M_t]}{i_\psi}}$$

$$\psi = \frac{b}{a} = 0.3$$

$$a \geq (4+1)3\sqrt{\left[\frac{0.7}{852.6}\right]^2 \times \frac{2.15 \times 10^3 \times 206.89 \times 10^3}{4 \times 0.3}}$$

$$a \geq 146.17 \text{ mm}$$
$$a = 150 \text{ mm}$$

7. Calculation of normal module:

$$M = \frac{2a}{(Z_1 + Z_2)} \times \cos\beta$$

$$= \frac{2 \times 150}{(20+80)} \times \cos 20°$$

$$= 2.8 \text{ mm}$$

The higher standard normal module is 3 mm.

$$k_{bl} = 0.7 \text{ for HB } 7350$$

$$k_\sigma = 1.5 \text{ for hardened steel.}$$

n = 2 − 5 for steel hardened.

$\sigma_{-1} = 0.35\,\sigma_u + 120$

$\sigma_u = 1550$ N/mm2

$\sigma_{-1} = 0.35 \times 1550 + 120$

$= 662.5$ N/mm²

$\sigma_b = 173.13$ N/mm²

To find $[\sigma_c]$:

$[\sigma_c] = C_R \times HRC \times k_{Cl}$

$= 26.5$ for alloy steel

MRC = 40 to 55

$k_{Cl} = 0.585$ for steel $H_B > 350$

$\sigma_C = 26.5 \times 55 \times 0.585$

$\sigma_1 = 852.6$ N/mm²

8. Revision of center distance:

$$a = \left(\frac{M_0}{\cos\beta} \right) \times \left(\frac{Z_1 + Z_2}{Z} \right)$$

$$= \frac{3}{\cos 20°} + \left(\frac{20 + 80}{2} \right) = 159.6 \text{ mm}$$

9. Calculation of b, d, v, and a_p:

Face width (b) = φ × u

$= 0.3 \times 159.6$

b = 42.8 mm

$$\text{Axial pitch} = \frac{\pi \times mn}{\sin\beta} = \frac{\pi \times 3}{\sin 20} = 27.5 \text{ mm}$$

$$\text{Pitch diameter of opinion}\,(d_1) = \frac{m}{\cos\beta} \times Z_1$$

$$= \frac{3}{\cos 20° \times 20}$$

$$= 63.85 \text{ mm}$$

$$\text{Pitchline velocity}\,(V) = \frac{\pi d_1 N_1}{60}$$

$$= \frac{\pi \times 63.85 \times 10^{-3} \times 1500}{60}$$

$$= 5.014 \ m/s$$

To find:

$$\varphi_P = \frac{s}{d_1} = \frac{47.8}{63.85} = 0.74$$

10. Calculation of value dimension of pinion and gear:

Module = 3 mm

Number of teeth Z_1 = 20 and Z_2 = 80

Pitch circle diameter, d_1 = 63.85 mm

$$d_2 = \frac{M}{\cos\beta} \times Z_2$$

$$= \frac{3}{\cos 20°} \times 80° = 255.4 \ mm$$

Center distance a = 159.6 mm

Height factor f_0 = 1

Bottom clearance c = 0.25 m = 0.25 × 3 = 0.75 mm

Tooth depth h = 2.25m = 2.25 × 3 = 0.75 mm

$$\text{Tip diameter,}\,d_{a1} = \left(\frac{Z_1}{\cos\beta} + 2\,f_0 \right) \times m$$

$$= \left[\frac{20}{\cos 20°} + 2 \times 1 \right] \times 3$$

$$= 69.85 \ mm$$

$$d_{a2} = \left(\frac{Z_2}{\cos\beta} + 2\,f_0 \right) \times m$$

$$= \left[\frac{80}{\cos 20°} + 2 \times 1 \right] \times 3$$

$$= 261.40 \ mm$$

2.6   Pressure Angle in the Normal and Transverse Plane

Pressure angle is the angle between a tangent to the tooth profile and a line perpendicular to the pitch surface.

Pressure angle is the angle at a pitch point between the line of pressure which is normal to the tooth surface and the plane tangent to the pitch surface. The pressure angle, as defined in this catalog, refers to the angle when the gears are mounted on their standard center distances.

Boston Gear manufactures both 14-1/2° and 20° PA, involute, full depth system gears. While 20°PA is generally recognized as having higher load carrying capacity, 14-1/2°PA gears have extensive use.

The lower pressure angle results in less change in backlash due to center distance variation and concentricity errors. It also provides a higher contact ratio and consequent smoother, quieter operation provided that undercut of teeth is not present.

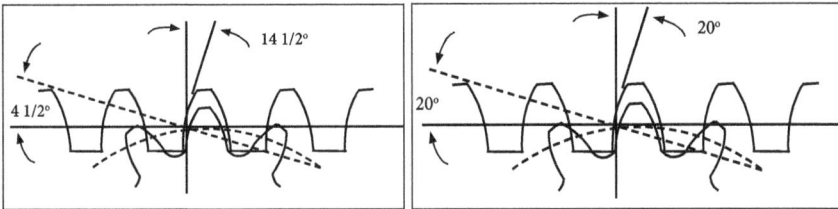

2.6.1 Equivalent Number of Teeth

The equivalent number of teeth (also called virtual number of teeth), Zv, is defined as the number of teeth in a gear of radius Re:

$$Z_v = \frac{2R_e}{m_n} = \frac{d}{m_n \cos^2 \psi}$$

2.6.2 Forces for Helical Gears

In designing a gear, it is important to analyze the magnitude and direction of the forces acting upon the gear teeth, shafts, bearings, etc. In analyzing these forces, an idealized assumption is made that the tooth forces are acting upon the central part of the tooth flank.

Below table represents the equations for tangential (circumferential) force F_t (kgf), axial (thrust) force F_x (kgf), and radial force F_r in relation to the transmission force F_n acting upon the central part of the tooth flank. T and T1 shown therein represent input torque (kgf·m).

The three-dimensional view of the forces acting on a helical gear tooth is shown in the figure. Resolving F_n.

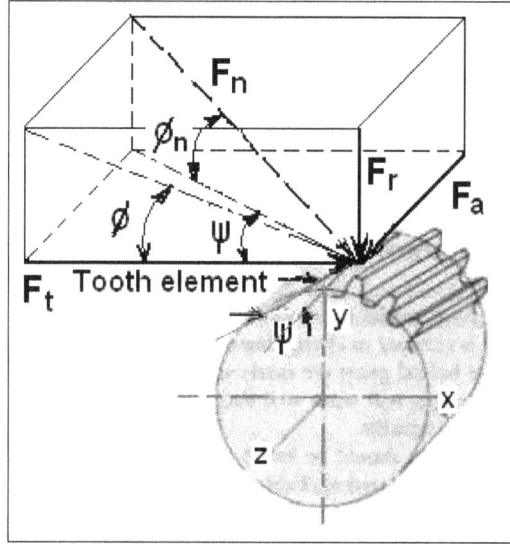

Tooth Force Acting on a Right Hand Helical Gear.

$$F_r = F_n \sin \emptyset_n$$
$$F_t = F_n \cos \emptyset_n \cos \psi$$
$$F_a = F_n \cos \emptyset_n \sin \psi$$
$$F_r = F_t \tan \emptyset$$
$$F_a = F_t \tan \psi$$
$$F_n = \frac{F_t}{\cos\varphi_n \cos\psi}$$

Figure illustrates the tooth forces acting on spur and helical gears. For spur gears, the total tooth force consists of components tangential F_t and radial F_r forces. For helical gears, component F_a is added and normal section NN is needed to show a true view of total tooth force Fn.

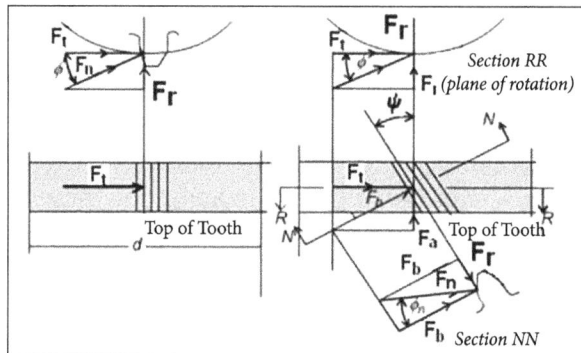

Comparison of Force Components on Spur and Helical Gears.

The vector sum F_t and F_a is labeled F_b; the subscript b being chosen because F_b is the bending force on the helical tooth (just as Ft is bending force on the spur tooth).

The force component associated with power transmission is only Ft.

$$F_t = 10000 \; w/v.$$

Where, F_t is in (N), W is in kW, and V is the pitch line velocity in (m/s).

$$F_b = F_t \,/ \cos \psi$$
$$F_r = F_b \; \tan \varnothing_n$$
$$F_r = F_t \; \tan \varnothing$$
$$\tan \varnothing_n = \tan \varnothing \cos \psi$$

Types of gears		F_t:Tangential force	F_x : Axial force		F_r : Radial force	
Spur gear		$F_t = \dfrac{2000T}{d}$	—		$F_t \tan \alpha$	
Helical gear			$F_t \tan \beta$		$F_t \dfrac{\tan\alpha_n}{\cos\beta}$	
Straight bevel gear		$F_t = \dfrac{2000T}{d_m}$	$F_t \tan\alpha \sin\delta$		$F_t \tan\alpha \cos\delta$	
Spiral bevel gear		d_m is the central reference diameter $d_m = d - b \sin\delta$	When convex surface is working:			
			$\dfrac{F_t}{\cos\beta_m}(\tan\alpha_n \sin\delta - \sin\beta_m \cos\delta)$		$\dfrac{F_t}{\cos\beta_m}(\tan\alpha_n \cos\delta + \sin\beta_m \sin\delta)$	
			When concave surface is working:			
			$\dfrac{F_t}{\cos\beta_m}(\tan\alpha_n \sin\delta + \sin\beta_m \cos\delta)$		$\dfrac{F_t}{\cos\beta_m}(\tan\alpha_n \cos\delta - \sin\beta_m \sin\delta)$	
Worm gear pair	Worm (Driver)	$F_t = \dfrac{2000T_1}{d_1}$	$F_t \dfrac{\cos\alpha_n \cos\gamma - \mu \sin\gamma}{\cos\alpha_n \sin\gamma + \mu \cos\gamma}$		$F_t \dfrac{\sin\alpha_n}{\cos\alpha_n \sin\gamma + \mu \cos\gamma}$	
	Worm Wheel (Driven)	$F_t \dfrac{\cos\alpha_n \cos\gamma - \mu \sin\gamma}{\cos\alpha_n \sin\gamma + \mu \cos\gamma}$	F_t			
Screw gear $\left(\begin{array}{c}\Sigma = 90° \\ \beta = 45°\end{array}\right)$	Driver gear	$F_t = \dfrac{2000T_1}{d_1}$	$F_t \dfrac{\cos\alpha_n \sin\beta - \mu \cos\beta}{\cos\alpha_n \cos\beta + \mu \sin\beta}$		$F_t \dfrac{\sin\alpha_n}{\cos\alpha_n \cos\beta + \mu \sin\beta}$	
	Driven gear	$F_t \dfrac{\cos\alpha_n \sin\beta - \mu \cos\beta}{\cos\alpha_n \cos\beta + \mu \sin\beta}$	F_t			

Bevel, Worm and Cross Helical Gears

3.1 Straight Bevel Gear

Bevel Gear Uses

The bevel gears used for transmitting power at a constant velocity ratio between two shafts whose axes intersect at a certain angle.

Virtual Number of Teeth

An imaginary spur gear considered in a plane perpendicular to the tooth of the bevel gear at the larger end is known as virtual spur gear.

The number of teeth Z_v on this imaginary spur gear is called virtual number of teeth in bevel gears:

$$Z_v = \frac{2}{\cos \delta}$$

Where, Z = Actual number of teeth on the level gear and δ = Pitch angle.

Problems

1. Let us Design a pair of cast iron bevel gears for a special purpose machine tool to transmit 3.5 kW from a shaft at 500 rpm to another at 800 rpm. The gears overhang in their shafts. Life required is 8000 hours.

Given data:

- P=3.5 kW
- N1 = 800rpw
- N2 = 500rpm
- α = 20°
- h_{fe} = 8000 hours

Solution

The material of pistons gear are same, we have to work (or) design for pinion only:

1. Gear ratio:

$$i = \frac{N_1}{N_2} = \frac{800}{500} = 1.6$$

Pitch angle:

For right angle level gear:

$\tan \delta 2 = i = 1.6$

$\delta 2 = \tan{-1} (1.6)$

$\delta 2 = 58°$

$\delta 1 = 90° - \delta 2 \Rightarrow 32°$

2. Material for pinion and gear:

Cast iron, Grade 35 heat treated

$\sigma_u = 350 \text{ N/mm}^2$

3. Gear life in hours = 8000 hours.

Gear life in cycle N = 8000 × 800 × 60

$N = 3.84 \times 108$ Cycles

4. Calculation of initial design torque (Mt):

$$[M_t] = M_t \times K \times K_d \times K_o$$

$$M_t = \frac{60 \times P}{2\pi N_1} \Rightarrow \frac{60 \times 3.5 \times 10^3}{2\pi \times 800} = \frac{2.1 \times 10^4}{2\pi \times 800} = 41.77 \text{ N} - \text{mm}$$

$K \cdot K_o = 1.3$ and $K_d = 1.25$ [assume]

$[M_t] = 41.77 \times 1.3 \times 1.25$

$[M_t] = 67.89 \text{ N} - \text{m}$

5. Calculation of E_{eq} and $|\sigma_b|$ and $|\sigma_c|$:

To find:

$E_{eq} = 1.4 \times 105 \text{ N/mm}^2$ for cast iron

$\sigma_u = 280 \text{ N/mm}^2$

$$[\sigma_b] = \frac{1.4 K_{bL}}{n \cdot K_\sigma} \times \sigma_{-1}$$

$$K_{bc} = 9\sqrt{\frac{10^7}{N}} = 9\sqrt{\frac{10^7}{3.84 \times 10^8}} = 0.607 \text{ for C.I.}$$

$K_\sigma = 1.2$ for C.I. from 5.15

$n = 2$ from 5.15 and $\sigma_{-1} = 0.45\sigma u$

$\sigma_u = 350$ N/mm² for C.I Table 5.3

$\sigma^{-1} = 0.45 \times 350 = 157.5$ N/mm²

$$[\sigma_b] = \frac{1.4 \times 0.667}{2 \times 1.2} \times 157.5$$

$$[\sigma_b] = 61.28 \text{ N/mm}^2$$

6. To find $[\sigma_c]$:

We know, that design contact stress:

$$[\sigma_c] = C_B \times H_b \times K_{CL}$$

$C_B = 2.3$, table 5.18

$H_b = 200$ to 260, table 5.18

$$K_{Cl} = 6\sqrt{\frac{10^7}{N}} \Rightarrow 6\sqrt{\frac{10^7}{3.84 \times 10^8}} = 0.544 \text{ N/mm}^2$$

$$[\sigma_c] = 2.3 \times 260 \times 0.544 = 325.57 \text{ N/mm}^2$$

7. Calculation of cone distance (R):

$$R \geq \psi_y \sqrt{i^2 + 1} \ \sqrt[3]{\left[\frac{0.72}{(\psi_y - 0.5)[\sigma_c]}\right]^2 \times \frac{E_{eq}[M_t]}{i}}$$

$$\psi_y = \frac{R}{b} = 3 \ (\text{assumed})$$

$$\therefore R \geq 3\sqrt{(1.6)^2 + 1}.$$

$$\sqrt[3]{\left[\frac{0.72}{(3 - 0.5)325.57}\right]^2 \times \frac{1.4 \times 10^5 \times 41.7 \times 10^3}{1.6}}$$

(or)

R = 47 mm

Assume Z_1 = 20 mm; Then Z_2 = i × Z1 = 1.6 × 20 = 32

Virtual number of teeth:

$$Z_{v1} = \frac{Z_1}{\cos \delta_1} = \frac{20}{\cos 32°} = 23.78$$

$$Z_{v2} = \frac{Z_2}{\cos \delta_2} = \frac{40}{\cos 58°} = 75.48$$

8. Calculation of transverse module (m_t):

$$m_t = \frac{R}{0.5\sqrt{Z_1^2 + Z_2^2}} = \frac{47}{0.5\sqrt{20^2 + 40^2}} = 2.28 \text{ mm}$$

9. Revision of cone distance (r):

We know,

$$R = 0.5 \, m_t \sqrt{Z_1^2 + Z_2^2} \Rightarrow 0.5 \times 2.5\sqrt{20^2 + 40^2} = 55.9 \text{ mm}$$

10. Calculation of b, m_{av}, d_{jav}, V and ψ_y:

Face width (b):

$$b = \frac{R}{\psi_y} = \frac{55.9}{3} = 18.63 \text{ mm}$$

Average Module (m_{av}):

$$m_{av} = m_t - \frac{b \sin \delta_1}{Z_1}$$

$$= 2.5 - \frac{18.63 \sin 32}{20}$$

$$= 2.006 \text{ mm}$$

Average ped of pinion (d_{lay}):

$$d_{lay} = m_{av} \times Z_1$$

$$d_{lay} = 40.12 \text{ m}$$

Pitch line velocity (V):

$$V = \frac{\pi \times d_{1av} \times N_1}{60} = \frac{\pi \times 40.12 \times 10^{-3} \times 800}{60}$$
$$= 1.68 \text{ m/s}$$

To Find:

$$\psi_y = \frac{b}{d_{1av}}$$
$$= \frac{18.63}{40.12} = 0.46$$

11. S.I Quality 6 bevel gears are assumed.

12. Revision of design torque:

We know,

$$[M_t] = M_t \times K \times K_d \times K_o$$

Where,

K = 1.1, for b/draw

K_d= 1.35, for IS quality 6 and U up to 3 m/s

K_o = 1.25, for medium shock

$[M_t]$ = 41.77 × 1.1 × 1.35 × 1.25

$[M_t]$ = 86.15 N−m

13. Check for bending:

$$\sigma_b = \frac{R\sqrt{i^2 + 1}[M_t]}{(R - 0.5b)^2 \times b \times m_t \times y_{vl}}$$

Where, $y_{vl} \approx 0.408$ for Z_{vl} = 23.78

$$\sigma_b = \frac{55.9\sqrt{2^2 + 1}[86.15 \times \omega^3]}{(2 - 0.5 \times 18.63) \times 18.63 \times 2.5 \times 0.408}$$
$$= 27.97 \times 606 \text{ N/mm}^2$$

We find σ_b, $[\sigma_b]$. Then, the design is not safety.

Trial-2:

Now, we will try with increased transverse module 3 mm. Repeating from step-9 again.

$$R = 0.5 \times m_t \times \sqrt{Z_1^2 + Z_2^2}$$

$$= 0.5 \times 3 \times \sqrt{20^2 + 40^2} = 67.08 \text{ mm}$$

$$b = \frac{R}{\psi_y} = \frac{67.08}{3} = 22.36 \text{ mm}$$

$$m_{av} = m_t - \frac{b \sin \delta_1}{Z_1} = 3 - \frac{22.36 \sin 32°}{20}$$

$$= 2.40 \text{ mm} \approx 25 \text{ mm}$$

$$d_{1av} = m_{av} \times Z_1 = 2.40 \times 20 = 50 \text{ mm}$$

$$V = \frac{\pi \times d_{1av} \times N_1}{60} \Rightarrow 2.094 \text{ m/s}$$

$$\psi_y = \frac{b}{d_{1av}} = \frac{22.36}{50} = 0.447$$

IS quality 6 bevel gear is assumed:

$K = 1.1$

$K_d = 1.35$

$K_o = 1.25$

$[M_t] = Mt \times K \times K_d \times K_o$

$\quad = 33.24 \text{ N-m}$

$$\sigma_b = \frac{67.08 \sqrt{2^2 + 1} \times 33.24 \times 10^3}{\left((67.08) - 0.5 \times 22.36\right)^2 \times 22.36 \times 3 \times 0.408}$$

$$= 58.3 \text{ N/mm}^2$$

Now found σ_b, $< [\sigma_b]$. The design is safe.

14. Check for wear strength:

$$\sigma_c = \frac{0.72}{(R - 0.56)} \left[\frac{\sqrt{(i^2 + 1)^3}}{i \times b} \times E_{eq} [M_t] \right]^{1/2}$$

$$= \frac{0.72}{(67.08-0.5 \times 22.36)} \left[\frac{\sqrt{(2^2+1)^3}}{2 \times 22.36} \times 1.4 \times 10^5 \times 33.24 \times 10^3 \right]^{1/2}$$

$$= 439.303 \text{ N} / \text{mm}^2$$

We find $\sigma_c < [\sigma_c]$. Thus the design is safe.

15. Calculation for basic dimensions of pinion and gears:

- Transverse module = m_t = 3 mm.

- No. of teeth Z_1 = 20; Z_2 = 40.

- Pitch circle diameter $d_1 = m_t = Z_1$ = 60 mm.

- $d_2 = m_t \times Z_2$ = 120 mm.

- Cone distance R = 67.08 mm.

- Face width b = 22.36 mm.

- Pitch angle = $\delta_1 = 32°$ and $\delta_2 = 58°$.

- Tip diameter $d_{a1} = m \in (Z_2 + 2\cos\delta 1)$.

- d_{a1} = 125.08 mm.

- $d_{a2} = m_t (Z_2 + 2\cos\delta_2)$.

- d_{a2} = 123.17 mm.

- Height factor = f_o = 1.

- Clearance C = 0.2.

Addendum angle:

$$\tan\theta_{a1} = \tan\theta_{a2} = \frac{m_t \times f_o}{R}$$

$$= \frac{3 \times 1}{67.08} \Rightarrow 0.0447$$

$$\theta_{a1} = \theta_{a2} = 2 \cdot 56°$$

Deddendum angle:

$$\tan\theta_{f1} = \tan\theta_{f2} = \frac{m_t (F_o + C)}{R} = \frac{3(1+0.2)}{67.08} = 0.05366$$

$$\theta_{f_1} = \theta_{f_2} = 3.07°$$

Tip angle $\delta_{a_1} = \delta_1 + \theta_{a_1} = 34.56°$.

$$\delta_{a_2} = \delta_2 + \theta_{a_2} = 58° + 2.56° = 60.56°.$$

Root angle $\theta_{f_1} = \delta_1 - \theta_{f_1} = 32 - 3.07 = 28.93°$.

$$\theta_{f_2} = \delta_2 - \theta_{f_2} = 58 - 3.07 = 54.93°.$$

Virtual number of teeth $Z_{V_1} = 23$; $Z_{V_2} = 90$.

2. Design a bevel gear drive to transmit 7.36 kw at 1440 rpm for the following data. Gear ratio = 3. Material for pinion and gear C45 surface hardened.

Given data:

      Design, Bevel gear Power = 7.3 kw

      N = 1440 rpm

      Speed ratio = 3

      Material type = C45 surface hardened

Solution

1. Material selection:

C45 surface hardened to 55RC core hardness 350 BHN.

$$\sigma_U = 720 \text{ N/mm}^2; \ \sigma_Y = 360 \text{ N/mm}^2$$

2. Assume life as 10,000 hours:

$$10,000 \times 60 \times 1440 \text{ rpm} = 86.4 \times 10^7 \text{ cycles}$$

3. Design stress calculation:

$$[\sigma_c] = C_R \ \text{HRC} \cdot \text{kcl}$$

$$= 23 \times 55 \times 0.585$$

$$= 740 \text{ N/mm}^2$$

From data book:

Here $C_R = 23$

HRC = 55

kcl = 0.585

$$[\sigma_b] = \frac{1.4 \times k_{b1}}{n \cdot k_r} \sigma_{-1}$$

$$= \frac{1.4 \times 1 \times 320}{2.5 \times 1.5}$$

$$= 119.5 \text{ N/mm}^2$$

Where,

$$\sigma_{-1} = 0.25 \left(\sigma_U + \sigma_y \right) + 50$$

$$= 0.25 (720 + 360) + 50$$

$$\sigma_{-1} = 320 \text{ N/mm}^2$$

$$k_{b1} = 1, \; ; n = 2.5 \; ; k_\sigma = 1.5$$

4. Determination of cone distance (R):

$$R \geq \psi_y \sqrt{i^2 + 1} \; 3\sqrt{\left[\frac{0.72}{(\psi_r - 0.5)(\sigma_c)} \right]^2 \times \frac{E \times [M_t]}{i}}$$

Assume $\psi_Y = \dfrac{R}{b} = 3$

$$i = 3$$

$$[M_t] = k_o \; k.k_d \cdot M_t$$

k = 1 (Assumed)

$$k.k_d = v_3 (\text{Assumed})$$

$$M_t = \frac{7.36 \times 10^3 \times 60}{2\pi \times 1440} = 48.8 \text{ Nm}$$

$$[Mt] = k_o \; k.k_d \; M_t = 1 \times 1.3 \times 43.10 = 63.5 \text{ N} - \text{m}$$

$$E = 2.1 \times 10^5 \text{ N/mm}^2$$

$$R \geq 3\sqrt{3^2 + 1}$$

$$3\sqrt{\left(\frac{0.72}{(3 - 0.5 \times 7.40)}\right)^2 \times \frac{2.1 \times 10^5 \times 63.5 \times 10^3}{3}}$$

$$R \geq 83.8 \text{ mm}$$

5. Determination of M_t:

$$R = 0.5 \, M_t Z_1 \sqrt{i^2 + 1}$$
$$83.8 = 0.5 \, M_t (18)\sqrt{3^2 + 1}$$

Assume $Z_1 = 1.8$

$$M_t = 2.9 \text{ mm} \, \Omega \, 3 \text{ mm}$$

6. Revice R and b:

$$R = 0.5 \times 3 \times 18\sqrt{3^2 + 1} = 85.38 \text{ mm}$$

$$b = R/3 = 85.38/3 = 23.46 \text{ mm}$$

$$\sigma_b = \frac{R\sqrt{i^2 + 1}[M_t]}{(R - 0.56)^2 \, b \, M_t \, Y_v}$$

Where, $Z_{eq} = \dfrac{Z_1}{\cos \delta_1} = 18/\cos 13.5 = 19$

$$Y_v = 0.383$$

$$\sigma_b = \frac{85.35 \times \sqrt{(3^2 + 1)(109.4 \times 10^3)}}{(85.3 - 0.5 \times 28.46)^2 \times (23.4^2 \times 3 \times 0.383)}$$

$$= 178 \text{ N/mm}^2 \, [\sigma_b] = 49 \text{ N/mm}^2$$

Here, the induced stress is more than the material value change the three dimension to get the safe values.

Assume $M_t = 4 \text{ mm}$

$$R = 0.5 \, M_t Z_1 \sqrt{C^2 + 1} = 0.5 \times 4 \times 15 \times \sqrt{10} = 114 \text{ mm}$$

$$b = 114/3 = 38 \text{ mm}$$

$$M_{ax} = M_t - \frac{b \sin 1}{Z_1} = 4 - \frac{38 \sin 18.5}{8} = 3.3 \text{ mm}$$

$$a_{1av} = Z_1 M_{av} = 18 \times 3.3 = 59.4 \text{ mm}$$

$$V = \frac{\pi d_{1av} n_1}{60 \times 1000} = \frac{\pi \times 544 \times 1440}{60 \times 1000} = 4.5 \text{ m/s}$$

$$k_d = 1.4 \; ; k = 1.6$$

Since, k and k_d values same, $[M_t]$ remains unaltered.

$$[M_t] = 109.4 \times 10^3 \text{ N-mm}$$

Revice k, k_d, and $[M_t]$:

$$M_{ax} = M_t - \frac{b \sin \sigma_1}{Z_1}$$

$$\tan \delta_2 = i = 3$$

$$\delta_2 = \tan^{-1}(3) = 71.5$$

$$\delta_1 = 90° - 71.5˘ = 18.5°$$

So, $$M_{av} = 3 - \frac{28.46 \sin 18.5}{18} = 2.498 \text{ mm}$$

$$V = \text{Pitch line velocity} = \frac{\pi d_{1av} a_1}{60 \times 1000} \text{ m/sec}$$

$$d_{1av} = Z_1 M_{av} = 18 \times 2.498 = 44.96 \text{ mm}$$

$$V = \frac{\pi \times 44.96 \times 1440}{60 \times 1000} = 3.4 \text{ M/S.}$$

∴ Select (1.4) from table.

(Is – quality of '6' bevel gear is assumed).

$$k = 1.6 \; ; \text{For } b/d_{1av} = 23.4/44.9 = 0.64$$

$$[M_t] = k_o k k_d M_t = 1 \times 1.6 \times 1.4 \times 48.80 \times 10^3$$

$$[M_t] = 109.4 \times 10^3 \text{ N-mm}$$

Calculation of Induced Stress:

$$\sigma_c = \frac{0.72}{(R-0.5\,b)}\left[\frac{\sqrt{\left(i^2+1\right)^3}\ E_x\left[M_t\right]}{i\,b}\right]^{1/2}$$

$$= \frac{0.72}{85.38-0.5\times28.46}\times\left[\frac{\sqrt{\left(3^2+1\right)^3}\left(2.1\times10^5\times109.4\times10^3\right)}{3\times28.46}\right]^{1/2}$$

$$= 944\ N/mm^2\left[\sigma_c\right]=740\ N/mm^2$$

$$\text{So, } \sigma_c = \frac{0.72}{114-0.5\times38}\times\left[\frac{\sqrt{\left(3^2+1\right)^3}\times2.1\times10^5\times109.4\times10^3}{3\times38}\right]^{1/2}$$

$$= 612\ N/mm^2\left[\sigma_c\right]=740\ N/mm^2 \text{ design is safe}$$

$$\sigma_b = \frac{114\times\left(\sqrt{3^2+1}\right)\times109.4\times10^3}{\left(114-0.5\times33\right)^2\times30\times4\times0.383}=75\ N/mm^2$$

$$= 75\ N/mm^2\left[\sigma_b\right] 119.5\ N/mm^2 \text{ design is safe}$$

Result:

Material Type = C45

Pinion gear teeth (21) = 18

Gear number of teeth = (22) = 54

Power = 736 kW

Speed = 1440 rpm

R = 83 mm

3. Let us design a bevel gear drive to transmit 7.5 kW at 1500 rpm. Gear ratio is 3.5. Material for pinion and gear is C45 steel. Minimum number of teeth is to be 25.

Given:

p = 7.5 kW

N_1 = 1500 rpm

$i = 3.5$

C45 steel

$Z_1 = 25$

Solution

Step 1: $i = 3.5$ (given)

Step 2: given C45 steel

$$[\sigma_c] = 5000 \text{ kgf/cm}^2$$

$$[\sigma_b] = 1400 \text{ kgf/cm}^2$$

$$\sigma_{-1} = 2700 \text{ kgf/cm}^2$$

$$HB = 215$$

Step 3:

$$R \ge \psi_y \sqrt{i^2+1}\ 3\sqrt{\left(\frac{0.72}{(\psi_y-05)[\sigma_c]}\right)^2 \frac{E[M_t]}{i}} \qquad ...(1)$$

$$[M_t] = 1.3 \times 97420 \frac{kW}{n1}$$

Cone distance,

$= 633.23$ kgf/cm

$E = 2.15 \times 10^6$ kgf/cm^2

$i = 3.5,\ \psi_y = 3$

$(1) \Rightarrow R \ge 18.05$

$R = 20$ cm

Step 4:

Given $Z_1 = 25$

$Z_2 = 88$

Step 5:

Module,

$$m_{av} \geq 1.28 \; 3\sqrt{\frac{[M_t]}{y_v(\sigma_b)\psi_m Z_1}}$$

$$Z_v = \frac{Z_1}{\cos \delta_1}$$

$$\tan \delta_2 = i$$

$$\delta_2 = 740$$

$$\therefore \delta_1 = 16°$$

$$y_v = 0.427$$

Assume $\psi_m = 10$

$$\Rightarrow m_{av} \geq 0.207 \text{ cm}$$

$$m_{av} = 0.3 \text{ cm}$$

$$\therefore M_t = m_{av} + \frac{b}{Z_1}\sin \sigma_1$$

$$\psi_y = \frac{R}{b} \Rightarrow b = 6.66 \text{ cm}$$

$$\Rightarrow M_t = 0.37 \text{ cm}$$

Step 6:

$$\sigma_c = \frac{0.72}{(R-0.56)}\sqrt{\frac{(i^2+1)3}{ib}E[M_t]}$$

$$= 2292.38 \text{ kgf/cm}^2 [\sigma_c]$$

$$\sigma_b = \frac{R\sqrt{i^2+1}[M_t]}{(R-0.56)^2 \text{ bm } y_v \cos \alpha}$$

$$= 167.78 \text{ kgf/cm}^2 < [\sigma_b]$$

Step 7:

$$d_1 = mt Z_1 = 9,25 \text{ cm}$$

$$d_2 = mt\, Z_2 = 32.56 \text{ cm}$$

4. Let us design a cast from bevel gear drive for a pillar drilling machine to transmit 1.5 kW at 800 rpm to a spindle at 400 rpm. The gear is to work for 40 hours per week for 3 years. Pressure angle is 20°. Check the design and calculate the basic dimensions.

Given data:

$$P = 1.5 \text{ kW}$$

$$N_1 = 800 \text{ rpm}$$

$$N_2 = 400 \text{ rpm}$$

$$\alpha = 20°$$

To find:

i) Design a level gear drive.

Solution

Since the materials of pinion and gear are same, we have to design only the pinion.

$$i = \frac{N_1}{N_2} = \frac{800}{400} = 2.$$

1) Gear ratio:

Pitch angles:

For right angle bevel gears,

$$\tan \delta_2 = i = 2.$$

$$\text{or } \delta_2 = \tan^{-1}(2) = 63.43°.$$

$$\delta_1 = 90 - \delta_2 = 90° - 63.43 = 26.570.$$

2) Material for pinion and gear:

Cast iron, Grade 35 heat treated.

$$64 = 350 \text{ N/mm}^2.$$

3) Gear life:

In hours = (40 hrs/week) × (52 Weeks/Year × 3 years).

$$= 6240 \text{ hours.}$$

Gear life in Cycles, $N = 6240 \times 500 \times 60 = 29.52 \times 107$ Cycles.

4) Calculation of initial design torque $[M_t]$:

We know that $[M_t] = M_t \times k \times k_d \times k_o$

Where $M_t = \dfrac{60 \times P}{2\pi N_1} = \dfrac{60 \times 1.5 \times 10^3}{2\pi \times 800} = 17.905$ mm

$k \cdot k_o = 1.3$

$k_d = 1.25$, assuming, medium shock

$[M_t] = 17.905 \times 1.3 \times 1.25 = 29.095$

5) Calculation of E_q, $[\sigma_b]$ and $[\sigma_c]$:

To find $[E_q]$:

$E_q = 1.4 \times 10^5$ N/mm² for cast iron, $\sigma_4 > 280$ N/mm

To find $[\sigma_b]$:

We know that the design bending stress:

$[\sigma_b] = \dfrac{1.4 \; kb_1}{n \cdot k_t} \times \sigma_{-1}$, for rotation in one direction.

Where, $k_{b1} = 9\sqrt{\dfrac{10^7}{N}} = 9\sqrt{\dfrac{10^7}{29.952 \times 10^7}} = 0.8852$

for C.I from table 5.14

$k_\sigma = 1.2$ for C.I

$n = 2$,

$\sigma_{-1} = 0.45 \; \sigma_u$

But $\sigma_u = 350$ N/mm², for C.I

$\sigma_{-1} = 0.45 \times 350 = 157.3$ N/mm²

Then $[\sigma] = \dfrac{1.4 \times 0.8 \times 52}{2 \times 1.2} \times 157.5 = 81.33$ N/mm

To find $[\sigma_c]$:

We know that the design control stress,

$$\left[\sigma_c\right] = C_B \times H_B \times k_{cd}$$

$$C_B = 2.3$$

$$k_{cl} = 6\sqrt{\frac{10^7}{N}} = 6\sqrt{\frac{10^7}{29.952 \times 10^7}}$$

= 0.833 for C.I from tables 5.19

$$\left[\sigma_c\right] = 2.3 \times 260 \times 0.833 = 498.05 \text{ N/mm}^2$$

6) Calculation of cone distance (R):

We know, that:

$$R \geq \psi_y \sqrt{i^2 + 1} \sqrt{\left[\frac{0.72}{\left(\psi_y - 0.5\right)\left[\sigma_c\right]}\right]^2 \times \frac{E_q\left[M_t\right]}{1}}$$

where $\psi_4 = R/b = 3$, Initially assumed.

$$\therefore R \geq 3\sqrt{2^2 + 1} \; 3\sqrt{\left[\frac{0.72}{\left(3 - 0.5\right)498.08}\right]^2 \times \frac{1.4 \times 10^5 \times 17.905 \times 10^3}{2}}$$

≥ 50.3

$R = 51 \text{ mm}$

7) Assume $Z_1 = 20$, Then $Z_2 = 1 \times Z_1 = 2 \times 20 = 40$

Virtual number of teeth:

$$ZV_1 = \frac{Z_1}{\cos \delta_1} = \frac{20}{\cos 26.57} \approx 23$$

$$ZV_2 = \frac{Z_2}{\cos \delta_2} = \frac{40}{\cos 63.43} \approx 90$$

8) Calculation of transverse module (M_t):

We know that,

$$= \frac{R}{0.5\sqrt{Z_1^2 + Z_2^2}} = \frac{51}{0.5\sqrt{20^2 + 40^2}} = 2.28 \text{ mm}$$

The nearest higher standard transverse module is 2.5 mm.

9) Revision of cone distance (R):

We know that, $R = 0.5 \ M_t \sqrt{Z_1^2 + Z_2^2}$

$$= 0.5 \times 2.5 \sqrt{20^2 + 40^2} = 55.9 \text{ mm}$$

10) Calculation of b, M_{av}, d_{1av}, V and :

Face width (b):

$$b = \frac{R}{\phi_y} = \frac{55.9}{3} = 18.63 \text{ mm}.$$

Average module (Mav):

$$M_{av} = M_t - \frac{b \sin \delta_1}{Z_1}$$

$$= 2.5 - \frac{18.63 \times \sin 26.572}{20}$$

$$= 2.083 \text{ mm}$$

Average pcd of pinion:

$$\left(d_{1av} \right) : d_{1av} = m_{av} \times Z_1 = 2.083 \times 20 = 41.66 \text{ m}$$

Pitch line velocity (V):

$$V = \frac{\pi \times d_{1av} \times N_1}{60} - \frac{\pi \times 41.66 \times 10^{-3} \times 800}{60}$$

$$= 1.745 \text{ M/s}$$

To find:

$$\phi_y = \frac{b}{d_{1av}} = \frac{18.63}{41.66} = 0.447$$

11) Is quality 6 bevel gear is assumed.

12) Revision of design torque $[M_t]$:

We know that:

$$[M_t] = M_t \times k \times k_d \times k_o$$

Where,

$k = 1.1,\ for/d_{1av} \leq 1,$

$k_d = 1.25,$ for IS quality 6 and U up to 3 m/s.

$k_o = 1.23,$ for Medium shock.

$[M_t] = 17.905 \times 1.1 \times 1.35 \times 1.25 = 33.24$ N–m.

13) Check for bending:

We know that the induced bending stress,

$$\sigma_b = \frac{R\sqrt{i^2+1}\,[M_t]}{(R-0.56)^2 \times b \times m_t \times y_{v1}}$$

Where,

$Y_{v1} \approx 0.405,$ for $Z_{v1} = 23.$

$$\sigma_b = \frac{55.9\sqrt{2^2+1} \times 33.24 \times 10^3}{(55.9 - 0.5 \times 18.63)^2 \times 18.63 \times 0.408}$$

$= 100.75$ N/mm.

We find $\sigma_b < [\sigma_b]$. Thus, the design is not satisfactory.

Trial 2:

Now we will try with increased transverse module 3 mm.

$$R = 0.5 \times m_t \times \sqrt{Z_1^2 + Z_2^2} = 0.5 \times 3 \times \sqrt{20^2 + 40^2} = 67.08$$

$$b = \frac{R}{\psi_Y} = \frac{67.08}{3} = 22.36 \text{ mm}$$

$$M_{av} = M_t - \frac{b \sin \delta_1}{Z_1} = \frac{3 - 22.36 \times \sin 26.57°}{20} = 2.5 \text{ mm}$$

$$d_{1\ av} = M_{av} \times Z_1 = 2.5 \times 20 = 50 \text{ mm}$$

$$V = \frac{\pi \times d_{1\ av} \times N_1}{60} = \frac{\pi \times 50 \times 10^{-3} \times 800}{60} = 2.094 \text{ m/s}$$

$$\psi_Y = \frac{b}{d_{1av}} = \frac{22.36}{50} = 0.447.$$

IS quality 6 bevel gear is assumed.

$$k = 1.1$$

$$k_d = 1.35$$

$$k_o = 1.25$$

$$[M_t] = M_t \times k \times k_d \times k_o = 17.905 \times 1.1 \times 1.35 \times 1.25$$

$$= 33.24 \text{ N} - \text{m}$$

$$\sigma_b = \frac{67.08 \sqrt{2^2 + 1} \times 33.24 \times 10^3}{(67.08 - 0.5 \times 22.36)^2 \times 22.36 \times 3 \times 0.408} = 58.3 \text{ N/mm}^2$$

Now we find $\sigma_b < [\sigma_b]$. Thus, the design is satisfactory.

14) Check for wear strength:

We know that the induced contact stress:

$$\sigma_c = \frac{0.72}{(R - 0.5b)} \left[\frac{\sqrt{(i^2 + 1)^3}}{i \times b} \times E_{eq} [M_t] \right]^{1/2}$$

$$= \frac{0.72}{(67.08 - 0.5 \times 22.36)} \left[\frac{\sqrt{(2^2 + 1)^3}}{2 \times 22.36} \times 1.4 \times 105 \times 33.24 \times 103 \right]^{1/2}$$

$$= 439.33 \text{ N/mm}^2$$

We find $\sigma_c < [\sigma_c]$. Thus, the design is satisfactory.

15) Calculation of Basic dimensions of pinion and gear:

Transverse module:

$$M_t = 3 \text{ mm}$$

Number of teeth:

$$Z_1 = 20 \text{ and } Z_2 = 40$$

Pitch circle diameter:

$$d_1 = M_t \times Z_1 = 3 \times 20 = 60 \text{ mm}$$

$$d_2 = M_t \times Z_2 = 3 \times 40 = 120 \text{ mm}$$

Cone distance:

$$R = 67.08 \text{ mm}$$

Face width:

$$b = 22.36 \text{ mm}$$

Pitch angles:

$$\delta_1 = 26.57° \text{ and } \delta_2 = 63.43°$$

Tip diameter:

$$da_1 = M_t\left(Z_1 + 2 \cos \delta_1\right) = 3\left(20 + 2 \cos 26.57\right)$$

$$= 65.37 \text{ mm}$$

$$da_2 = M_t\left(Z_2 + 2 \cos \delta_2\right) = 3\left(40 + 2 \cos 63.43\right)$$

$$= 122.68 \text{ mm}$$

Height factor:

$$f_o = 1$$

Clearance:

$$C = 0.2$$

Addendum angle:

$$\tan \theta_{a1} = \tan \theta_{a2} = \frac{M_t \times f_o}{R} = \frac{3 \times 1}{67.08}$$

$$= 0.0447$$

$$\theta_{a1} = \theta_{a2} = 2.56°$$

Dedendum angle:

$$\tan \theta_{f1} = \tan \theta_{f2} = \frac{M_t\left(f_o + c\right)}{R} = \frac{3\left(1 + 0.2\right)}{67.08}$$

$$= 0.05366$$

$$\theta_{f1} = \theta_{f2} = 3.07°$$

$$\delta a_1 = \delta_1 + \theta_{a1} = 26.57° + 2.56° = 29.13°$$

$$\delta a_2 = \delta_2 + \theta a_2 = 62.43° + 2.56° = 65.99°.$$

Root angle:

$$\delta f_1 = \delta_1 - \theta f_1 = 26.57° - 3.07° = 23.5°$$

$$\delta f_2 = \delta_2 - \theta f_2 = 63.43° - 3.07 = 60.36°.$$

Virtual number of teeth:

$$ZV_1 = 23 \text{ and } ZV_2 = 90.$$

3.1.1 Tooth Terminology

Bevel Gear

Bevel gears transmit power between two intersecting shafts at any angle or between non-intersecting shafts. They are classified as straight and spiral tooth bevel and hypoid gears.

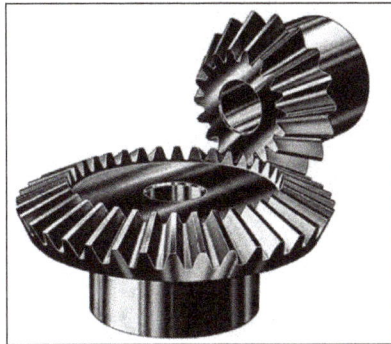

Bevel Gear in Mesh

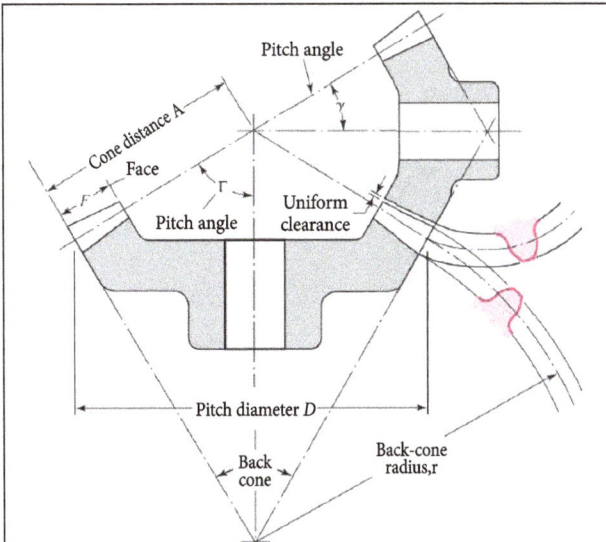

When intersecting shafts are connected by gears, the pitch cones (analogous to the pitch cylinders of spur and helical gears)are tangent along an element, with their apexes at the intersection of the shafts as in Fig where two bevel gears are in mesh.

The size and shape of the teeth are defined at the large end, where they intersect the back cones. Pitch cone and back cone elements are perpendicular to each other. The tooth profiles resemble those of spur gears having pitch radii equal to the developed back cone radii r_{bg} and r_{bp} and are shown in Figure. Below which explains the nomenclatures of a bevel gear.

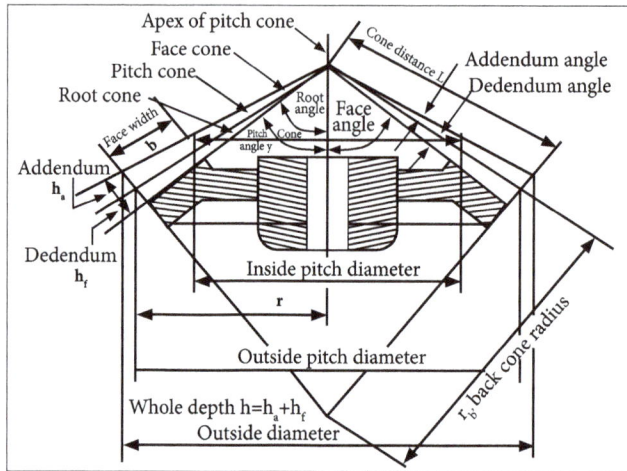

3.2   Tooth Forces and Stresses

In designing a gear, it is important to analyze the magnitude and direction of the forces acting upon the gear teeth, shafts, bearings, etc. In analyzing these forces, an idealized assumption is made that the tooth forces are acting upon the central part of the tooth flank.

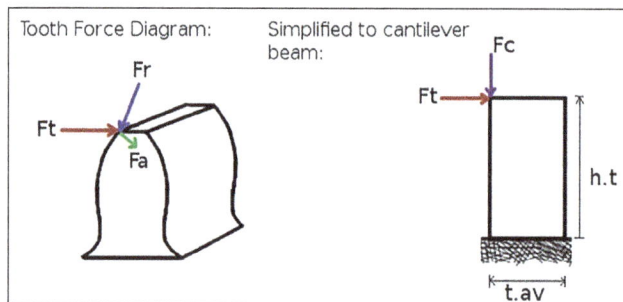

Below table represents the equations for tangential (circumferential) force F_t (kgf), axial (thrust) force F_x(kgf) and radial force F_r in relation to the transmission force Fn acting upon the central part of the tooth flank. T and T_1 shown, therein represent input torque (kgf·m).

Types of gears		F_t: Tangential force	F_x: Axial force	F_r: Radial force
Spur gear		$F_t = \dfrac{2000\,T}{d}$	-	$F_t \tan \alpha$
Helical gear			$F_t \tan \beta$	$F_t \dfrac{\tan \alpha_a}{\cos \beta}$
Straight bevel gear		$F_t = \dfrac{2000T}{d_m}$	$F_t \tan \alpha \sin \delta$	$F_t \tan \alpha \cos \delta$
Spiral bevel gear		d_m is the central reference diameter $d_m = d - b \sin \delta$	When convex surface is working: $\dfrac{F_t}{\cos \beta_m}\left(\tan \alpha_n \sin \delta - \sin \beta_m \cos \delta\right)$	$\dfrac{F_t}{\cos \beta_m}\left(\tan \alpha_n \cos \delta + \sin \beta_m \sin \delta\right)$
			When concave surface is working: $\dfrac{F_t}{\cos \beta_m}\left(\tan \alpha_n \sin \delta + \sin \beta_m \cos \delta\right)$	$\dfrac{F_t}{\cos \beta_m}\left(\tan \alpha_n \cos \delta - \sin \beta_m \sin \delta\right)$
Worm gear pair	Worm (driver)	$F_t = \dfrac{2000T_1}{d_1}$	$F_t = \dfrac{\cos \alpha_n \cos \gamma - \mu \sin \gamma}{\cos \alpha_n \cos \gamma + \mu \sin \gamma}$	
	Worm wheel (driven)	$F_t = \dfrac{\cos \alpha_n \cos \gamma - \mu \sin \gamma}{\cos \alpha_n \sin \gamma + \mu \cos \gamma}$	F_t	$F_t \dfrac{\sin \alpha_n}{\cos \alpha_n \sin \gamma + \mu \cos \gamma}$
Screw gear $\left(\begin{array}{c}\Sigma = 90° \\ \beta = 45°\end{array}\right)$	Driver gear	$F_t = \dfrac{2000T_1}{d_1}$	$F_t = \dfrac{\cos \alpha_n \sin \beta - \mu \cos \beta}{\cos \alpha_n \cos \beta + \mu \sin \beta}$	
	Driven gear	$F_t = \dfrac{\cos \alpha_n \sin \beta - \mu \cos \beta}{\cos \alpha_n \cos \beta + \mu \sin \beta}$	F_t	$F_t = \dfrac{\sin \alpha_n}{\cos \alpha_n \cos \beta + \mu \sin \beta}$

The pinion pitch diameter d_p and gear pitch diameter d_G can be found using the following equation.

$$P = \frac{N_G}{d_G} = \frac{N_P}{d_P}$$

Solving for the pitch diameters,

$$d \quad \underline{\quad\quad}$$

$$d_P = \frac{N_P}{P}$$

The tangential force can be found using the following torque equations. Where all variables are magnitudes only. See diagram of tooth forces for directions.

$$F_{t,P} = \frac{T_P}{r_{av,P}}$$

Where,

$$r_{av,P} = \left(d_p - F * \sin(\gamma)\right)/2$$

$$\tan(\gamma) = \frac{d_P}{d_G} = \frac{N_P}{N_G} = \frac{n_G}{n_P} = \frac{1}{m}$$

γ = Half cone angle of Pinion.

The tangential forces acting on the gear teeth are equal in magnitude and opposite in direction.

$$F_{t,G} = F_{t,P}$$

Where,

$F_{t,G}$ = Magnatude of the tangential force acting on the gear.

$F_{t,P}$ = Magnatide of the tangaental force acting on the pinion.

From tangential force, can calculate radial and axial forces,

$$F_{a,P} = F_{t,P} * \tan(\phi)\sin(\gamma)$$

$$F_{r,P} = F_{t,P} * \tan(\phi)\cos(\gamma)$$

To simplify for subsequent stress analysis, the radial and axial force can be resolved into a tooth compression force,

$$F_{c,P} = F_{r,P} \sin(\gamma) + F_{a,P} \cos(\gamma).$$

There is also a shearing component that acts along the tooth face, but for now will be neglected. If this component were to be calculated, it would be found by,

$$F_{s,P} = F_{r,P} \cos(\gamma) - F_{a,P} \sin(\gamma).$$

The diagram of tooth forces for direction of forces.

3.2.1 Equivalent Number of Teeth

The equivalent number of teeth (also called virtual number of teeth), Z_v, is defined as the number of teeth in a gear of radius R_e:

$$Z_v = \frac{2R_e}{m_n} = \frac{d}{m_n \cos^2 \psi}.$$

3.2.2 Estimating the Dimensions of Pair of Straight Bevel Gears

Problem

A design a bevel gear drive to transmit 3.5 kW. Speed ratio = 4. Driving shaft speed = 200 rpm. The drive is non-reversible. Pinion is of steel and wheel of Cl. Assume a life of 25,000 hours.

Given data:

Torque = 411 N-m

N = 475 rpm

Speed ratio = 9.

Solution

1. Selection of materials:

Alloy steel 40 Ni2 Crl Mo 28

Hardened and tempered, surface hardness 55 RC and core hardness 350 BHN. Material is same for both pinion and gear and hence only the pinion is designed.

2. Design stresses:

Life = N = 36000 × 60 × 475 = 102.6 × 10⁷ cycles.

Life = $N = 36000 \times 60 \times 475 = 102.6 \times 10^7$ cycles.

For this life, for the material mentioned in step (1),

$$[\sigma_b] = 173.1 \text{ N/mm}^2, \ [\sigma_c] = 852.6 \text{ N/mm}^2$$

$$a \geq (i+1) 3 \sqrt{\left[\frac{0.7}{\sigma_c}\right]^2 \frac{E[M_t]}{i\psi}}$$

3. Center distance calculation:

$$M_t = \text{pinion torque} = \frac{\text{gear torque}}{i}$$

$$= 411/4 = 102.76 \text{ Nm}$$

Assume, $k_o, k, k_d = 1.3$

$$[M_t] = 1 \times 1.3 \times 102.76 = 133.6 \text{ Nm}$$

Assume, $\psi = 0.5$, $E = 2.1 \times 10^5 \text{ N/mm}^2$

$$\geq (4+1)3 \sqrt{\left[\frac{0.7}{852.6}\right]^2 \frac{2.1 \times 10^5 \times 133.6 \times 10^3}{4 \times 0.5}}$$

$$\geq 106.5 \text{ mm}$$

Adopt standard center distance referring to RIO series (Appendix-1)

$$a = 125 \text{ mm}$$

4. Calculation of m_n, d_1, b, P_a and $[M_t]$:

Assume ß = helix angle = 10°.

$$Z_1 = 17, Z_2 = 4 \times 17 \times 68$$

$$m_n = \frac{2a \cos \beta}{z_1 + z_2} = \frac{2 \times 125 \times \cos 10°}{17 + 68} = 2.9 \text{ mm}$$

We should have a standard module. Change the helix angle or change the tooth numbers without introducing much deviation in the speed ratio.

Trial 2:

Try with $m_n = 2.5$. Take $Z_1 = 20, Z_2 = 80$.

$$2.5 = \frac{2 \times 125 \times \cos \beta}{(20 + 80)}$$

i. e., ß = 0 (The helix angle becomes zero. It becomes a spur gear. Not satisfactory. For helical gears ß = 8° to 25°).

Trial 3:

$$Z_1 = 20, Z_2 = 79$$

$$2.5 = \frac{2 \times 125 \times \cos \beta}{(20 + 79)}$$

i.e., ß = 8.1°(acceptable).

Re-calculate speed ratio, i = 79/20 = 3.95.

$$d_1 = \frac{m_n z_1}{\cos \beta} = \frac{2.5 \times 20}{\cos 8.1°} = 50.5 \text{ mm}$$

$$d_2 = \frac{m_n z_2}{\cos \beta} = \frac{2.5 \times 79}{\cos 8.1°} = 199.49 \text{ mm}$$

$$b = \psi a = 0.5 \times 125 = 62.5 \text{ mm}$$

$$p_a = \text{axial pitch} = \frac{\pi m_n}{\sin \beta} = \frac{\pi \times 2.5}{\sin 8.1°} = 55.7 \text{ mm}$$

We find, $b > p_a$.

Pitch line velocity:

$$V = \frac{\pi d_1 n_1}{60 \times 1000} = \frac{\pi \times 50.5 \times 475}{60 \times 1000} = 1.26 \text{ m/s}$$

$$\psi_p = \frac{b}{d_1} = \frac{62.5}{50.5} = 1.2$$

Assuming class 8 gears, we find $k_d = 1.1$ for 1.26 m/s.

$$k = 1.14 \text{ for } \psi_p = 1.2$$

Now, $kk_d = 1.14 \times 1.1 = 1.254$

Assumed $kk_d = 1.3 >$ actual kk_d.

Assumed kk_d is marginally greater than the actual kk_d.

Hence, initial $[M_t]$ it-self can be used for calculating induced stresses and these stresses will be slightly greater than the actual induced stresses.

5. Calculation of induced stresses:

Check for σ_c : $[M_t]$ has not been modified from the initial value. We have increased a considerably and hence there is no need to check the induced contact stress.

Check for σ_b :

$$\sigma_b = 0.7 \frac{i+1}{a \, m_n \, b \, y_v} [M_t]$$

$$z_{eq} = \frac{z_1}{\cos^3 \beta} = \frac{20}{\cos^3 8.1°} = 21.$$

$$y_v = 0.395 (\text{for 21 teeth})$$

$$= \frac{0.7 \times (3.95 + 1) \times 133.6 \times 10^3}{125 \times 2.5 \times 62.5 \times 0.395}$$

$$= 60.6 \ \text{N/mm}^2 \left[\sigma_b\right] = 173.1 \ \text{N/mm}^2.$$

The design is satisfactory.

3.3 Worm Gear Merits and Demerits

Worm Gear

Worm gears are used for transmitting power between two non-parallel, non-intersecting shafts. High gear ratios of 200:1 can be got.

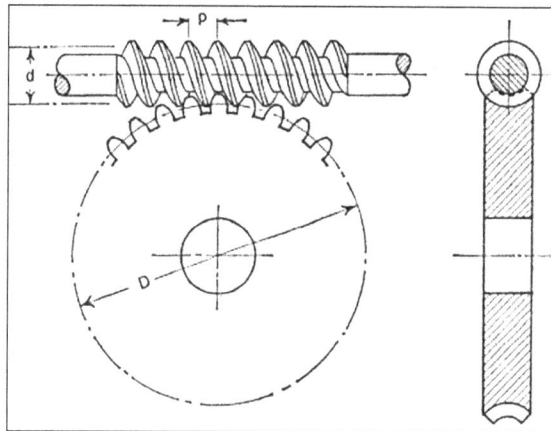

Nomenclature of Worm Gear.

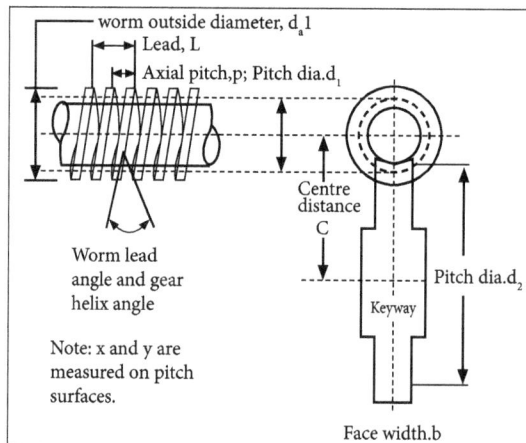

Note: x and y are measured on pitch surfaces.

The geometry of a worm is similar to that of a power screw. Rotation of the worm simulates a linearly advancing involute rack.

The geometry of a worm gear is similar to that of a helical gear, except that the teeth are curved to envelop the worm.

Enveloping the gear gives a greater area of contact but requires extremely precise mounting.

As with a spur or helical gear, the pitch diameter of a worm gear is related to its circular pitch and number of teeth Z by the formula,

$$d_2 = \frac{Z_2 p}{\pi}.$$

When the angle is 90° between the nonintersecting shafts, the worm lead angle λ is equal to the gear helix angle Ψ. Angles λ and Ψ have the same hand.

The pitch diameter of a worm is not a function of its number of threads, Z_1.

This means that the velocity ratio of a worm gear set is determined by the ratio of gear teeth to worm threads; it is not equal to the ratio of gear and worm diameters.

$$\frac{\omega_1}{\omega_2} = \frac{Z_2}{Z_1}.$$

Worm gears usually have at least 24 teeth and the number of gear teeth plus worm threads should be more than 40:

$$Z_1 + Z_2 > 40.$$

A worm of any pitch diameter can be made with any number of threads and any axial pitch.

Integral worms cut directly on the shaft can, of course, have a smaller diameter than that of shell worms, which are made separately.

Shell worms are bored to slip over the shaft and are driven by splines, key, or pin.

Strength considerations seldom permit a shell worm to have a pitch diameter less than,

$$d_1 = 2.4p + 1.1$$

The face width of the gear should not exceed half the worm outside diameter.

$$b \leq 0.5\, d_{a1}$$

Advantages of Worm Gear Drive:

- The worm gear drives can be used for speed ratio as high as 300:1.

- The operation is smooth and silent.

- The worm gear drives are compact compared with equivalent spur or helical gears for the same speed reduction.

- The worm gear drives are irreversible. It means that the motion cannot be transmitted from worm wheel to the work. This property of irreversible is advantages in load hoisting application like cranes and lifts.

3.4 Thermal Capacity, Stresses and Efficiency

Thermal Capacity

The continuous rated capacity of a worm gear set is often limited by the ability of the housing to dissipate friction heat without developing excessive gear and lubricant temperatures. Normally, oil temperature must not exceed about 200°F (93°C) for satisfactory operation. The fundamental relationship between temperature rise and rate of heat dissipation used for journal bearings does hold good for worm gearbox.

$$H = C_H \, A \left(T_o - T_a \right)$$

Where,

H – Time rate of heat dissipation (Nm/sec).

C_H – Heat transfer coefficient (Nm/sec/m² /°C).

A – Housing external surface area (m²).

T_o – Oil temperature (°C).

T_a – Ambiant air temperature (°C).

Surface area of A for conventional housing designs may be roughly estimated from the Equation.

$$A = 14.75 \, C^{1.7}$$

Where A is in m2 and C (the distance between the shafts) is in m. Housing surface area can be made far greater than the above equation value by incorporating cooling fins.

Rough estimates of C can be taken from the following figure.

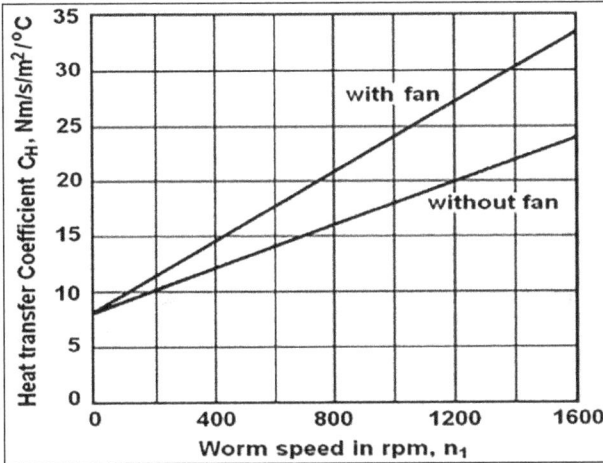

Influence of Worm Speed on Heat Transfer.

Efficiency η is the ratio of work out to work in. For, the usual case of the worm serving as input member.

$$\eta = \frac{F_{2t}V_{2t}}{F_{1t}V_1} = \frac{\cos\phi_n \cos\lambda - f\sin\lambda}{\cos\phi_n \cos\lambda + f\cos\lambda}\tan\lambda$$

$$\eta = \frac{\cos\phi_n - f\tan\lambda}{\cos\phi_n + f\tan\lambda}$$

The overall efficiency of a worm gear is a little lower because of friction losses in the bearings and shaft, seals and because of "churning" of the lubricating oil.

Problems

A 2 kW power is applied to a worm shaft at 720 mm. The worm is of quadruple start with 50 mm as pitch circle diameter. The worm gear has 40 teeth with 5 mm module. The pressure angle in the diametric plane is 20°. Let us determine (i) the lead angle of the worm, (ii) velocity ratio and (iii) center distance. And also, calculate efficiency of the worm gear drive and power lost in friction.

Given:

$P = 2$ KW; $n_1 = 720$ rpm; $d = 50$ mm;

$Z = 40$; $m = 5$ mm; $\alpha = 20°$.

Solution

$$q = \frac{d}{m} = 10$$

Lead angle of worm, $\gamma = \tan^{-1}\left(\dfrac{Z}{q}\right)$.

Assume number of stacks, $Z = 3$

$\qquad \therefore \gamma = 16.69°$

$\qquad i = \dfrac{z}{Z} = \dfrac{40}{3} = 13.33$

Center distance,

$$a = \left(\dfrac{z}{q}+1\right)\sqrt[3]{\left[\dfrac{540}{\dfrac{z}{q}[\sigma_c]}\right]^2} \; [M_t].$$

$$[M_t] = \dfrac{1.3 \times 97420 \times 2}{720} = 351.79 \text{ kgf cm}$$

$$[\sigma_c] = 5000 \text{ kgf/cm}^2$$

$\qquad \therefore a = 3.176 \text{ cm}$

$\qquad a = 10 \text{ cm}$

Efficiency of the drive,

Sliding velocity, $V = \dfrac{\pi \times 50 \times 10^{-3} \times 720}{60}$

$\qquad V = 1.88 \text{ m/s}$

$\qquad \Rightarrow \mu = 0.04$

$\qquad \therefore \mu = \tan^{-1} \mu = 2.29°$

$\qquad \therefore \mu = \dfrac{\tan \gamma}{\tan(\gamma + \rho)} = 84.72 \%$

Power lost due to friction,

$\qquad H_g = (1 - 2) P$

$\qquad = 1.69 \text{ KW}$

3.4.1 Estimating the Size of the Worm Gear Pair

The worm gear pairs are made up of a worm and a worm wheel. These gears allow for a very large reduction ratio in a single pair. They are very quiet and provide smooth transmission of power. Due to the situation of being a friction drive mechanism, worm gear pairs have very poor efficiency. Similar to helical gears, worm wheels are also cylindrical disks which have involute shaped teeth cut into their face at an angle. For worm gear pairs, this is the lead angle and it is identical to the angle cut into the worm.

The reduction ratio of a worm gear pair is developed as the ratio of the number of teeth on the worm wheel. The number of starts on the worm. As with helical gears, worm gear pairs must have the same module, pressure angle and direction of lead angle in order to mesh. Our worm gear pairs are offered in many materials, modules, numbers of teeth and reduction ratios.

All of the worm wheels that we offer allow for secondary operations such as opening the bore, adding of keyways, adding of tapped holes and the reduction of the hub diameter to be performed. Most of the worms that we offer allow for secondary operations such as opening the bore, adding of keyways, adding of tapped holes, and the reduction of the hub diameter or shaft diameter to be performed.

Problem

Design a worm gear drive to transmit 22.5 kW at a worm speed of 1440 rpm. Velocity ratio is 24:1. An efficiency of at least 85% is desired. The temperature rise should be restricted to 40°C. Let us determine the required cooling area.

Given data:

$p = 3.5 \text{ kW}$

Speed ratio = 4

$N = 200 \text{ rpm}$

Assume life = 25000 hrs.

Solution

1. Select material, pinion C45 surface hardened to 45 RC and core hardness 350 BHN.

$\sigma_u = 700 \text{ N/nim}^2$

Wheel: Ci grade 30,

$\sigma_y = 360 \text{ N/mm}$

$$\sigma_u = 300 \text{ N/mm}^2$$

2. Calculation of design stresses:

$$[\sigma_c]_p = C_R \text{ HRC } K_{cl}$$

$$= 23 \times 45 \times 0.585$$

$$= 605.5 \text{ N/mm}^2$$

$$C_R = 23$$

$$\text{HRC} = 45$$

$$K_{cl} = 0.585$$

$$[\sigma_b]_p = \frac{1.4 \ K_{b1}}{n \ K_\sigma} \sigma_{-1} \ \left(\text{for N} = 25000 \times 60 \times 200\right)$$

$$= 30 \times 107 \text{ cycles})$$

$$= \frac{1.4 \times 1 \times 315}{2.5 \times 1.5}$$

$$= 117.6 \text{ N/mm}^2$$

$$\sigma = 0.25 \left(\sigma_n + \sigma_y\right) + 50$$

$$= 315 \text{ N/mm}^2$$

$$[\sigma_c]_w = C_B \cdot \text{HB} \cdot K_{cl}$$

$$= 2.3 \times 220 \times 0.714$$

$$= 361.3 \text{ N/mm}^2$$

$$K_{b1} = 1 \left(\text{for } 30 \times 10^7 \text{ cycles}\right)$$

$$K_\sigma = 1.5$$

$$n = 2.5$$

$$[\sigma_b]_w = \frac{1.4 \ K_{b1}}{n \ K_\sigma} \sigma_{-1}$$

$$N_w = (30/i) \times 10^7 = (30/4) \times 10^7$$

$$= 7.5 \times 10^7 \text{ cycles}$$

$$K_{cl} = 6\sqrt{\frac{10^7}{N_w}}$$

$$= 6\sqrt{\frac{10^7}{7.5 \times 10^7}} = 0.714$$

$$C_B = 2.3$$

HB = 220 (assumed)

$$\sigma_{-1} = 0.45$$

$$\sigma_u = 0.45 \times 300$$

$$= 135 \text{ N/mm}^2$$

$$K_{b1} = 9\sqrt{\frac{10^7}{N_w}}$$

$$= 9\sqrt{\frac{10^7}{7.5 \times 10^7}} = 0.799$$

$$= \frac{1.4 \times 0.799}{2.5 \times 1.2} \times 135$$

$$= 50.3 \text{ N/mm}$$

$$K_\sigma = 1.2$$

n = 2.5 (no heat treatment)

3. Determination of R (Design the pinion).

$$R \geq \psi_y \sqrt{i^2+1} \; 3\sqrt{\left(\frac{0.72}{(\psi_y - 0.5)[\sigma_c]}\right)^2 \frac{E[M_t]}{i}}$$

i = 4 (given)

CI grade 30 has σ_U 280 N/mm².

Hence, equivalent Young's modulus.

$$E = 1.7 \times 10^5 \text{ N/mm}^2$$

Assume, $\psi_y = 3$

$$[M_t] = k_o \cdot kk_d \cdot M_t$$

M_t = pinion torque.

$k_o = 1$ (assumed).

$kk_d = 1.3$ (assumed).

$$M_t = \frac{3.5 \times 10^3 \times 60}{2\pi \times 200} = 167.112 \text{ Nm}$$

$$[M_t] = 1 \times 1.3 \times 167.112 = 217.3 \text{ Nm}$$

$$R \geq 3\sqrt{16+1} \ 3\sqrt{\left(\frac{0.72}{(3-0.5)605.5}\right)^2 \frac{1.7 \times 10^5 \times 217.3 \times 10^3}{4}}$$

≥ 158.13 mm.

4. Determination of:

$$R = 0.5 \ m_t \sqrt{i^2 + 1} \cdot z_1$$

$158.13 - 0.5 \ m_t \sqrt{16+1} \cdot 18$ (assuming $z_1 = 18$).

$\therefore m_t = 4.26$ mm Take $m_t = 5$(standard).

5. Revise R and find b:

$$R = 0.5 \times 5 \times \sqrt{16+1} \cdot 18 = 185.54 \text{ mm}.$$

b = R/3 = 185.54/3 = 61.8 mm.

6. Calculate m_{av}, d_{1av} and pitch line velocity:

$$m_{av} = m_t \left(\frac{R - 0.5b}{R}\right)$$

(Derived from the formula given for d_{1av} in

to get m_{av}, Equation can also be used.)

$$= 5 \times \left(\frac{185.54 - 0.5 \times 61.8}{185.54} \right) = 4.17 \text{ mm}$$

$$\text{dlav} = z_1 m_{av} = 18 \times 4.17 = 75 \text{ mm}$$

Pitch line velocity,

$$V = \frac{\pi d_{1av} n_1}{60 \times 1000} = \frac{\pi \times 75 \times 200}{60 \times 1000} = 0.8 \text{ m/s}$$

7. Revise k, k_d and $[M_t]$:

$$b / d_{1av} = 61.8/75 = 0.824$$

Surface hardness is less than 350 BHN for wheel.

$$\therefore \ k = 1.1$$

$$k_d = 1, \text{ for class 6 gears, for } V = 1 \text{ m/s}$$

$$\left[M_t \right] = 1 \times 1.1 \times 1 \times 167.1 = 183.8 \text{ Nm}$$

8. Calculation of induced stresses

i) Stresses in pinion:

$$\sigma_c = \left(\frac{0.72}{R - 0.5b} \right) \left[\frac{\sqrt{(i^2 + 1)^3}}{ib} \cdot E[M_t] \right]^{1/2}$$

$$= \left(\frac{0.72}{185.54 - 0.5 \times 61.8} \right) \left[\frac{\sqrt{(4^2 + 1)^3}}{4 \times 61.8} \times 1.7 \times 10^5 \times 183.8 \times 10^3 \right]^{1/2}$$

$$= 438.4 \text{ N/mm}^2 \left[\sigma_c \right] = 605.5 \text{ N/mm}^2$$

$$\sigma_b = \frac{R\sqrt{i^2 + 1}[M_t]}{(R - 0.5 \ b)^2 bm_t y_v}$$

$$\tan \delta_2 = i = 4$$

$$\delta_2 = \tan^{-1} 4 = 76°$$

$$\delta_1 = 90° - 76° = 14°$$

$$z_{eq} = z./\cos \delta_1 = 18/\cos 14° = 19$$

$$\therefore y_v = 0.383$$

$$= \frac{185.54 \sqrt{16+1} \times 183.8 \times 10^3}{(185.54 - 0.5 \times 61.8)^2 \times 61.8 \times 5 \times 0.383}$$

$$= 49.7 \text{ N/mm}^2 \left[\sigma_b\right] = 117.6 \text{ N/mm1}$$

Design of the pinion is satisfactory.

ii) Check the stresses in the wheel:

$$\sigma_{cw} = \sigma_{cp} = 438.4 \text{ N/mm}^2 > \left[\sigma_c\right]_w = 361.3 \text{ N/mm}^2$$

Wheel does not have adequate wear strength.

Use the following relation to calculate the induced bending stress.

$$\sigma_{bw} \cdot y_w = \sigma_{bp} \cdot y_p$$

where, y_w is the form factor for the virtual number of teeth of the wheel.

$$z_2 = iz_1 = 4 \times 18 = 72$$

$$z_{eq} = Z_2/\cos \delta_2 = 72/\cos 76° = 297.6 = 300$$

$$y_v = 0.521$$

$$\sigma_{bw} \times 0.521 = 49.7 \times 0.383$$

$$\sigma_{bw} = 36.5 \text{ N/mm}^2 < \left[\sigma_b\right]_w = 50.3 \text{ N/mm}^2$$

Wheel has adequate beam strength.

Comments:

1. In order to increase the wear strength of the wheel, surface hardness may be raised by flame or induction hardening.

2. Taking $m_t = 6$, carry out the design calculations as trial 2.

Problems

1. Let us the Design a worm gear drive to transmit a power of 22.5 kW. The worm speed is 1440 rpm. and the speed of the wheel is 60 rpm. The drive should have a minimum efficiency of 80% and above. Select suitable materials for worm and wheel and decide upon the dimensions of the drive.

Soution

Given:

$$N_1 = 1440 \text{ rpm}; P = 22.5 \text{ KW};$$

$$N_2 = 60 \text{ rpm}; \eta_{derived} = 80\%$$

To find: Design the worm gear drive.

$$\text{Gear ratio} = \frac{1440}{60} = 24$$

1. Material selection:

Worm - Hardened Steel.

Worm Wheel - Phosphor bronze.

2. Selection of Z_1 and Z_2

For $\eta = 80$, $Z_1 = 3$

Then, $Z_2 = i \times Z_1 = 24 \times 3 = 72$

3. Calculation of q and r:

Diameter factor $q = \dfrac{d_1}{m_x} = 11 \ (\text{assumed})$

Lead angle: $\gamma = \tan^{-1}\left(\dfrac{Z_1}{q}\right) = \tan^{-1}\left(\dfrac{3}{11}\right) = 15.25°$

4. Calculation of F_t in terms of m_x:

Tangential load, $F_t = \dfrac{P}{V} + K_o$

$$V = \frac{\pi d_2 N_2}{60 \times 1000} = \frac{\pi \times (Z_2 \times m_x) \times N_2}{60 \times 1000}$$

$$= \frac{\pi \times 72 \times m_x \times 60}{60 \times 1000} = 0.2226 \ m_x m/s.$$

$K_o = 1.25$ assumed medium shock.

$$F_t = \frac{22.5 \times 10^3}{0.226 \ m_x} \times 1.25 = \frac{124446.9}{m_x}$$

5. Calculation of dynamic load (F_d):

$$F_d = \frac{F_t}{C_v}$$

$$C_v = \frac{6}{6+V}$$

$V = 5$ m/s is assumed.

$$= \frac{6}{6+5} = 0.545$$

$$F_d = \frac{124446.9}{m_x} \times \frac{1}{0.545}$$

$$= \frac{228342.9}{m_x}$$

6. Calculation of beam strength $\left(F_s\right)$:

$$F_s = \pi \times m_x \times b \times [\sigma_b] \times y$$

$$b = 0.75 \ d_1$$

$$= 0.75 \times q \ m_x = 0.75 \times 11 \ m = 8.25 \ m_x$$

$[\sigma_b] = 80$ N/mm², from table

$y = 0.125$, assuming $a = 20°$

$$F_s = \pi \times m_x \times 8.25 \ m_x \times 80 \times 0.125 = 259.18 \ m_x^2$$

7. Calculation of axial module (m_x):

$$F_s \geq F_d$$

$$259.18 \ m_x^2 \geq \frac{228342.9}{m_x}$$

$$m_x \geq 9.5 \ mm$$

Nearest higher axial pitch is 10 mm.

8. Calculation of b, d_2 and v:

Face width $b = 8.25\, m_x = 8.25 \times 10 = 82.5$ mm.

Pitch diameter of worm wheel (d_2):

$$d_2 = Z_2 \times m_x = 72 \times 10$$

$$= 720 \text{ mm.}$$

Pitch line velocity of worm Wheel (V):

$$0.226\, m_x = 0.226 \times 10 = 2.26 \text{ m/s.}$$

9. Recalculation of beam strength (F_s):

$$F_s = 259.18\, m_x^2 = 259.18 \times (10)^2$$

$$= 25918 \text{N.}$$

10. Recalculation of dynamic load (F_d):

$$F_d = \frac{F_t}{C_V}$$

$$C_V = \frac{6}{6+V} = \frac{6}{6+2.26} = 0.72$$

$$F_t = \frac{124446.9}{m_x} = \frac{124446.9}{10} = 12444.6 \text{ N.}$$

$$F_d = \frac{124446}{10} = 1244.46 \text{ N.}$$

11. Check for beam strength:

$F_d < F_s$. It means that the gear tooth has adequate beam strength and will not fail by breakage. Thus the. design is satisfactory.

12. Calculation of maximum wear load (F_w):

$$F_w = d_2 \times b \times K_W$$

$$K_w = 0.56 \text{ N/mm}^2 \text{ from table.}$$

$$F_w = 720 \times 82.5 \times 0.56 = 33264 \text{ N.}$$

13. Check for wear:

$F_d < F_w$. It means that the gear tooth has adequate wear capacity and will not wear out. Thus the design is safe and satisfactory.

14. Check for efficiency:

$$\eta_{actual} = 0.95 \frac{\tan \gamma}{\tan(\gamma + \rho)}$$

$$\rho = \text{Friction angle} = \tan^{-1}\mu$$

$$= \tan^{-1}(0.03) = 1.7°$$

$$\eta = 0.95 \times \frac{\tan 15.25°}{\tan(15.25 + 1.7°)}$$

$$\because \mu = \tan P$$

$$\mu = 0.03 \text{ assumed}$$

$$= 0.8498 \text{ or } 84.98\%$$

Thus the design is satisfactory.

2. Let us the design a worm gear drive with a standard center distance to transmit 7.5 kW from a worm rotating at 1440 rpm to a worm wheel at 20 rpm.

Given Data:

$$N_1 = 1440 \text{ rpm}$$

$$P = 7.5 \text{ kW}$$

$$N_2 = 20 \text{ rpm}$$

Assume $\eta = 82\%$

To Find:

Design the worm gear drive.

Solution

$$i = \frac{1440}{60} = 24$$

Gear ratio required:

1) Metal selection:

Worm - Hardened steel

Worm wheel - Phosphor bronze.

2) Selection of Z_1 and Z_2:

For $\eta = 82\%$ $Z_1 = 3$

Then $Z_2 = i \times Z_1 \Rightarrow T_2$.

3) Calculation of q and r3:

$$\text{diameter factor } q = \frac{d_1}{m_x} = 11$$

Lead angle:

$$W = \tan^{-1}\left(\frac{Z_1}{q}\right) = \tan^{-1}\left(\frac{3}{11}\right)$$

$$= 15.25°$$

4) Calculation of F_t in term of M_x:

$$\text{Tangential load } F_t = \frac{P}{V} \times K_o$$

$$V = \frac{\pi d_2 N_2}{60 \times 1000} \Rightarrow \frac{\pi(Z_2 \times m_x) \times N_2}{60 \times 1000}$$

$$= \frac{\pi + (72 \times mx) \times 20}{60 \times 50}.$$

$V = 0.075$ mx m/s.

$K_o = 1.25$ assuming medium shock

$$F_t = \frac{7.5 \times 10^3}{0.075\ m_x} = 1.25$$

$$F_t = \frac{125000}{m_x}.$$

5) Calculation of Dynamic Load:

$$\text{Dynamic load } F_d = \frac{F_t}{C_V}$$

$$C_V = \frac{6}{6+V}$$

Where, V = 5 m/s assume.

$$C_V = 0.545$$

$$F_d = \frac{125000}{m_x} \times \frac{1}{0.545} = \frac{229357.79}{m_x}$$

6) Calculation of beam strength (F) in terms of axial module:

$$F_s = \pi \times m_x \times b \times [\sigma_b] \times y$$

$$b = 0.75\, d_1$$

$$= 0.75 \times qm_x \Rightarrow 0.75 \times 11\, m_x \Rightarrow 8.25\, m_x$$

$$[\sigma_b] = 80\ \text{N/mm}^2$$

$$y = 0.125,\ \text{Assume}\ \alpha = 20°$$

$$F_s = \pi \times m_x \times 8.25\, m_x \times 80 \times 0.125$$

$$F_s = 259.18\ m^2$$

7) Calculation of a× IAL Module (m_x):

We know $F_s \geq F_d$

$$259.18\ m_x^2 \geq \frac{229357.79}{m_x}$$

$$m_x \geq 15.67$$

The nearest higher standard axial pitch is 16mm.

8) Calculation of (b), d_2 and V:

Face width (b):

$$b = 8.25\, m_x$$

$$b = 8.25 \times 16$$

$$b = 132\ \text{mm}$$

Pitch diameter, $(d_2) = Z_2 \times m_x \Rightarrow 72 \times 16$

$\qquad d_2 = 1152$ mm

Pitch line velocity of work wheel (V) = $0.226 \, m_x$

$\qquad V = 3.616$ m/s

9) Re-calculation of beam strength (F_s):

$\qquad F_s = 259.18 \, m_x^2$

$\qquad F_s = 259.18 \times (16)2$

$\qquad F_s = 66350.08$ N

10) Re-calculation of dynamic load (F_d):

$$F_d = \frac{F_t}{C_v}$$

$$C_v = \frac{6}{6+V} \Rightarrow \frac{6}{6+3.61} \Rightarrow 0.62$$

$$F_t = \frac{125000}{m_x} \Rightarrow \frac{125000}{16}$$

$\qquad F_t = 7812.5$ N

$$F_d = \frac{7812.5}{0.666}$$

$\qquad F_d = 12600.80$ N

11) Check for beam strength:

We find $F_d < F_s$. It means that the gear tooth has adequate beam strength and will not fail by breakage. The design is safe.

12) Calculation of Max Wear Load (F_w):

We know,

$$\eta_{actual} = 0.95 \, \frac{\tan W}{\tan (W+P)}$$

$$\rho = \text{Friction angle} = \tan^{-1}\mu\left[\mu = \tan\rho\right]$$

$$= \tan^{-1}(0.03) = 1.7° \dots \left[\mu = 0.03 \text{ assume}\right]$$

$$\eta = 0.95 \times \frac{\tan 15.25°}{\tan(15.25° + 1.7°)}$$

$$\eta = 84.9\%$$

We found that actual efficiency is greater than the desired efficiency. Thus, design is safe.

15) Calculation of basic dimensions of work and worm gear:

Axial load m_x = 8 mm.

No. of starts $Z_1 = 3$.

No. of teeth on work wheel: $Z_2 = 72$

Face width of worm wheel, b = 132 mm,

length of worm, $L \geq (12.5 + 0.09\, Z_2)m_x$

$$\geq (12.5 + 0.09(72)16$$

$$L = 30.6 \text{ mm}$$

Centre distance:

$$a = 0.5\, m_x\left(q + Z_2\right)$$

$$a = 664 \text{ mm}$$

Height factor $f_o = 1$

Bottom clearance:

$$C = 0.25 m_x$$

$$C = 4 \text{ mm}$$

Pitch diameter:

$$d_1 = q \times m = 176 \text{ mm}$$

$$d_2 = Z_2 \times m_x = 1152 \text{ mm}$$

Tip diameter:

$$d_{a1} = d_1 + 2f_o\, m_x = 176 + 2 \times 16$$

$$d_{a1} = 2848 \text{ mm}$$

$$d_{a2} = (Z_2 + 2f_0)m_x$$

$$d_{a2} = 1184 \text{ mm.}$$

Root distance:

$$df_1 = d_1 - 2f_0 \cdot m_x - 2C$$

$$= (176 - 2) \times 16 - 8$$

$$df_1 = 2776 \text{ mm}$$

$$df_2 = (Z_2 - 2f_0)m_x - 2C$$

$$= (72 - 2) \times 16 - 8$$

$$df_2 = 1112 \text{ mm.}$$

3. Let us the design a worm gear drive and determine the power loss by heat generation. The hardened steel worm rotates at 1500 rpm and transmits 10 kW to a phosphor bronze gear with gear ratio of 16.

Given:

Steel worm.

$$N_1 = 1500 \text{ rpm.}$$

$$P = 10 \text{ kW.}$$

Phosphor bronze wheel.

$$i = 16.$$

Solution

Step 1:

Given material,

Worm: Steel,

Wheel: Phosphor bronze,

Assume, V = 3 m/s.

Step 2:

$$a = \left(\frac{Z}{q} + 1\right) 3 \sqrt{\left[\frac{540}{\frac{Z}{q}[\sigma_c]}\right]^2} [M_t]$$

Centre distance,

Assume Z = 2

$$Z = iZ = 32$$

Initially choose $q = 11$

$$[\sigma_c] = 1590 \text{ kgf/cm}^2$$

$$[M_t] = 1 \times M_t = 1 \times \frac{P \times 60}{2\pi n_2}$$

$$i = \frac{n_1}{n_2} \Rightarrow n_2 = 94 \text{ rpm}$$

$$\therefore [M_t] = 10158.8 \text{ kgf cm}$$

$$(1) \Rightarrow a = 20.22 \text{ cm} = 22\text{cm}$$

Step 3:

Axial module,

$$m_x = 1.24 \ \sqrt[3]{\frac{[M_t]}{Z_q \, y_v \, [\sigma_b]}}$$

$$Z_v = \frac{Z}{\cos^3 \gamma}$$

$$\gamma = \tan^{-1}\left(\frac{Z}{q}\right) = 10.3°$$

$$Z_v = 34$$

$$y_v = 0.452 \ ; \ [\sigma_b] = 780 \text{ kgf/cm}^2$$

$$\therefore m_x = 0.538,$$

$$m_x = 0.6 \text{ cm}$$

Step 4:

$$q_c = \frac{540}{\left(\dfrac{z}{q}\right)} \sqrt{\left[\frac{\dfrac{Z}{q}+1}{a}\right]^3} \; [M_t]$$

$$= 1401.3 \; \text{kgf/cm}^2 \; [\sigma_c]$$

$$\sigma_b = \frac{1.9 \, [M_t]}{m_x^3 \, q \, Z \, y_v}$$

$$= 561.64 \; \text{kgf/cm}^2 \, [\sigma_c]$$

Step 5:

Face width of wheel, for $Z = 2$, $b = 0.75 \, d_1$

$$= 0.75 \, (qm_x)$$

$$= 4.95 \; \text{cm}$$

Length of the work,

$$L \geq (11 + 0.06 \, Z) \, m_x$$

$$L \geq 7.75$$

$$L = 9 \; \text{cm}.$$

3.5   Cross Helical Gear

The crossed helical gears are a type of Helical Gears only. They are the typical example of two helical gears applied for transmitting power between non-parallel and non-intersecting shafts. To operate a crossed helical gears perfectly, they should possess identical pressure angle and normal pitch. However, it is not essential to have a same helix angle or to be in opposite hand only. The Crossed Helical Gear is only recommended for a narrow range of applications, with relatively light loads and it is better to avoid it for precision meshes. What happens is that for the two gears contact takes place only in a point and not a line.

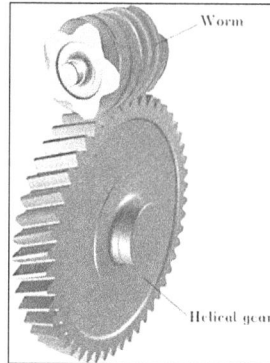

So the resulting high sliding loads taking place in between the teeth necessitates an extensive lubrication. Thus, it has been seen that very little power can actually be transmitted by using crossed helical gears. Cross helical gears are mounted on an intersecting shaft with the shaft angle at 90 degrees.

Manufacturing cross helical gears generally, the crossed helical gears are offered in four types of materials:

- * Steel * Chrome Steel * Cast aluminum * Bronze * Nylon.

Characteristics of cross helical gears:

- *Precision rating is poor * Skewed shafting * High sliding * Slow speeds * Only point contact * Lighter loads * Comparatively low velocity ratio * Principally used for allowing wide variety of speed ratios without changing center distance or gear size * The very good adaptability of crossed helical gears allows them to get over many of the difficulties in a speed-reducer design.

3.5.1 Helix Angles

A helix angle is that the angle between any helix and an axial line on its right, circular cylinder or cone. Common applications are screws, helical gears and worm gears.

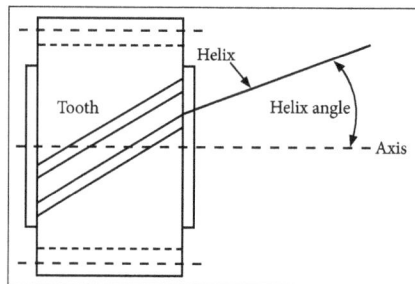

The helix angle references the axis of the cylinder, distinguishing it from the lead angle which references a line perpendicular to the axis. Naturally, the angle is that the geometric complement of the lead angle. The angle is measured in degrees.

3.5.2 Estimating the Size of the Pair of Cross Helical Gears

Helical Gears Connecting Non-Parallel Shafts Helical gears used to connect non-parallel shafts are commonly called spiral gears or crossed axis helical gears. If the shaft angle is 90 degrees, the gears will be of the same hand and the sum of the helix angles will be equal to the shaft angle (90 degrees).

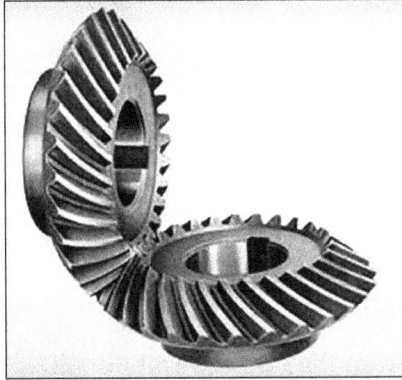

Helical gears used on non-parallel shafts must have the same normal pitch and normal pressure angles. They may however, be of the same or opposite hand depending on the shaft angle.

Problems

1. Let us the design a pair of helical gears to transmit 10 kw at 1000 rpm of the pinion. Reduction ratio of 5 is required.

Given data:

Helical gear.

> Power = 10 kw
>
> N = 1000 × PM
>
> i = 5.

Solution

Design procedure is same as that for spur gear; the following points are to be noted:

- Normal module should have a standing value.

- Form factor should be taken for virtual number of teeth.

- $Z_{eq} \geq 17$.

- Face width b > Pa for smooth operations.

1. Selection of Material:

Alloy steel 40N12Crl M028 hardened and tempered surface harness 55 RC and cone hardness 350 BHN.

$$\sigma_U = 1550 \text{ N/mm}^2$$

Material for pinion and gear are assumed to be same and hence only the pinion is designed.

$$[\sigma_b] = \frac{1.4 \times k_{b1}}{n \cdot k\sigma} \sigma_{-1}$$

2. Calculation of Design Stress:

$\sigma_7 = 0.35 \ \sigma_U + 120$ From PSG Design Data Book

$\sigma_{-1} = 662 \text{ N/mm2}$ For Alloy steel

n = 2.5

$k\sigma = 1.5$ Assume 10,000 hours of life.

N = Life = 10,000 × 60 × 1000

= 6 × 10^7 cycles

$k_{b1} = 0.7$ for cone hardness 150 BHN,

$$[\sigma_b] = \frac{1.4 \times 0.7 \times 665}{2.5 \times 1.5}$$

$$[\sigma_b] = 173.1 \text{ N/mm}^2$$

$$[\sigma_c] = C_R \cdot \text{HLC} \cdot \text{kclC}_R = 26.5$$

$$= 26.5 \times 55 \times 0.585 \text{ HRC} = 55$$

$$[\sigma_c] = 852.6 \text{ N/mm}^2 \text{kd} = 0.585$$

3. Calculation of central distance:

$$a \geq (i+1) \sqrt[3]{\left[\frac{0.7}{[\sigma_c]}\right]^3 \frac{E \cdot [M_t]}{i \psi}}$$

Where, $[M_t] = k_o k\, k_d \cdot M_t$

$$M_t = \frac{P \times 60}{2\pi \times N} = \frac{10 \times 10^3 \times 60}{2\pi \times 1000} = 95\ N-m.$$

Assume, $k_{kd} = 1.3$.

$$[M_t] = k_{kd} \cdot M_t = 1.3 \times 95 = 125\ N-m.$$

Assume, $\psi = 0.5$.

$$\in = 2 \times 10^5\ N/mm^2$$

$$a \ge (i+1) \cdot 3\sqrt{\left[\frac{0.7}{\sigma_c}\right]^2 \cdot \frac{\in \cdot [M_t]}{i\psi}}$$

$$\ge (5+1)\, 3\sqrt{\left(\frac{0.7}{852}\right)^2 \cdot \frac{2 \times 10^5 \times (125 \times 10^3)}{5 \times 0.5}}$$

$$> 115\ mm$$

$$a = 115\ mm.$$

Assume, $Z_1 = 20$ nos.

$$Z_2 = Z_1 = 5 \times 20 - 100\ nos.$$

Therefore, $M_n = \dfrac{2a \cdot \cos\beta}{Z_1 + Z_2} = \dfrac{2 \times 115\ \cos 15°}{(20 + 100)} = 1.85.$

$M_n = 3$ Standard module is selected.

Now revise a:

$$d_1 = \frac{M_n Z_1}{\cos\beta} = \frac{3 \times 20}{\cos 15°} = 62\ mm$$

$$d_2 = i\, d_1 = 5 \times 62 = 310\ mm$$

$$a = \frac{d_1\, f\, d_2}{2} = \frac{62 + 310}{2} = 186\ mm.$$

4. Calculation of b, P_G, $[M_t]$:

$$P_a = \frac{P_t}{\tan\beta} = \left[\frac{\pi d_1}{Z_1}\right] / \tan\beta = \left(\frac{T \times 62}{1020}\right) / \tan 15° = 36.3\ mm.$$

$$P_a = 36\ mm.$$

$$b = Pa = 0.5 \times 186 = 93\ mm.$$

Revice the $k_v k_d$ values, and $[M_t]$ values.

5. Calculation of induced stress:

σ_b induced bending stress; $Y_v = 2.402$ from data book.

$$\sigma_b = \frac{0.7 \times (i+1)[M_t]}{ab\ m_n\ Y_v} = \frac{0.7 \times (5+1)\left(\left[125 \times 10^3\right]\right)}{115 \times 93 \times 3 \times 0.402}$$

$$\sigma_b = 40.7\ \text{N/mm}^2 < [173]\text{N/mm}^2$$

Here, induced bending stress is less than design bending stress $[\sigma_b]$. So the design is safe.

σ_c Induced crushing stress:

$$\sigma_c = 0.7 \frac{(i+1)}{a} \sqrt{\frac{i+1}{i\ b}} \times E[M_t]$$

$$= 0.7 \times \frac{(5+1)}{115} \sqrt{\frac{(5+1)}{(5 \times 93)} \times 2 \times 10^5 \times 125 \times 10^3}$$

$$= 655\ \text{N/mm}^2 < [852]\ \text{N/mm}^2$$

Here, the induced crushing stress is less than design crushing stress so the design is safe.

Result:

Pair of Helical gear,

No. of teeth on pins $(Z_1) = 20$.

No. of teeth on gear $(Z_2) = 100$.

Pcd of Pinion $(d_1) = 62$ mm.

Pcd of Gear $d_2 = 186$ mm.

$N_1 = 1000$ rpm.

Power (P) = 10 kW.

Life = 10,000 hrs.

Speed ratio (P) = 5.

Center distance = 186 mm.

Face width = 93 mm.

$\psi = 0.5$.

Gear Boxes

4.1   Geometric Progression

The gearbox often referred as transmission is a unit that uses gears and gear trains to provide speed and torque conversions from a rotating power source to another device. Gearboxes are employed to convert input from a high speed power sources to low speed (Eg. Lift, Cranes and Crushing Machine) or into a many of speeds (Lathe, Milling Machine and Automobiles).

A gearbox that converts a high speed input into a single output it is called a single stage gearbox. It usually has two gears and shafts.

A gearbox that converts a high speed input into a number of different speed output it is called a multi-speed gear box. Multi speed gear box has more than two gears and shafts. A multi-speed gearbox reduces the speed in different stages.

4.1.1 Standard Step Ratio

In order to get a series of output speeds from a gearbox, geometric progression is used. By using geometric progression the speed is reduced uniformly in different stages. Geometric progression, also known as a geometric sequence is a sequence of numbers where each term after the first is found by multiplying the previous one by a fixed, non-zero number called the common ratio (called as progression ratio or step ratio in gear box design).

In gearbox design a set of preferred step ratio or preferred numbers is used to obtain the series of output speed of gearbox. The preferred step ratio is mentioned as basic series named as R5, R10, R20, R40 and R80. Each basic series has a specific step ratio.

4.1.2 Ray Diagram

A ray diagram is a representation of structural formula. It provides information such as speed in each stage, the transmission ratio in each stage. The total number of speeds and its values.

Difference between structural diagram and speed diagram:

Structural Diagram	Ray Diagram (Speed Diagram)
1) It gives information about number of gears available in each stage.	1) It gives information about ratio between two consecutive speeds.
2) $P_1(X_1)P_2(X_2)$	2) $3(1)\ 2(3)$

Ray Diagram of Gear Box

The ray diagram is a graphical representation of the data arrangement in general form. It is used to determine the specific values of all the transmission ratios and speeds of all the shafts in the drive.

Ray Diagram for 12 speed gear box

Draw the ray diagram for 12 speed gear box.

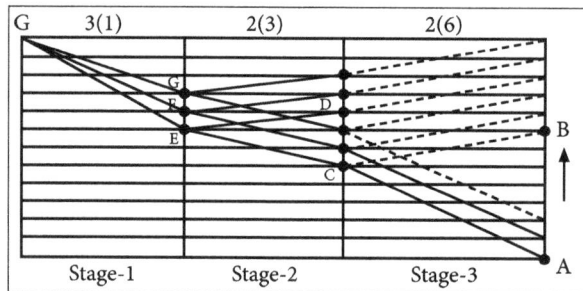

Problems

1. Let us a draw the ray diagram and kinematic lay out of a gear box for an all geared head stock of a lathe. The maximum and minimum speeds are to be 600 and 23 rpm respectively. Number of steps is 12 and drive is from a 3000 W electric motor running at 1440 rpm.

Given:

$n_{max} = 600$ rpm; $n_{min} = 23$ rpm

$n = 12$; driver; 3000 W, 1440 rpm

Solution

$$\phi^{n-1} = \frac{n_{max}}{n_{min}} \Rightarrow \phi = \left(\frac{600}{23}\right)^{\frac{1}{11}} = 1.345$$

The Speeds:

23, 31, 42, 56, 75, 101, 136, 183, 246, 331, 445, 598.

Structural Formula:

12 = 3(1)·	2(3)	2(6)
I	II	III
Stage	Stage	Stage.

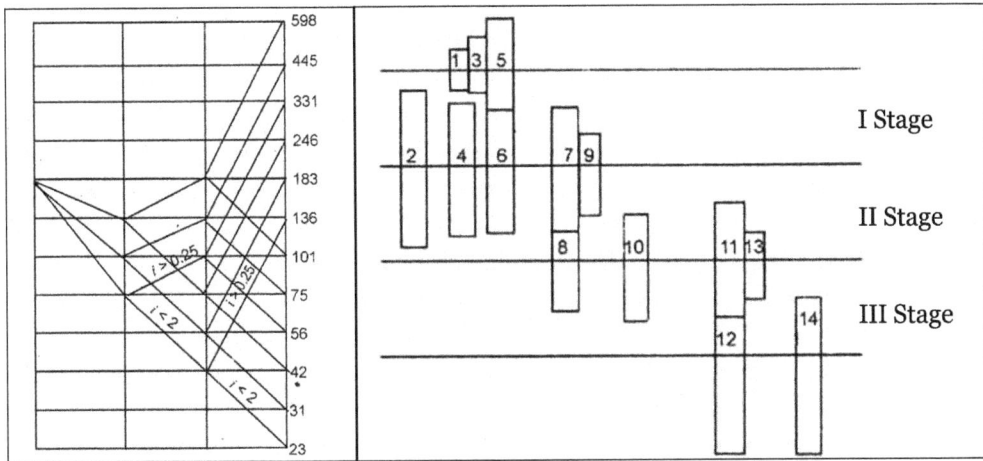

Kinematic Layout.

2. Let us a sketch three possible ray diagrams for a 6-speed gear box with 2×3 arrangement. Choose the best possible ray diagram. It is given the suitable explanation for the same.

Given data:

$$n = 6$$

$$N_{min} = 160 \text{ rpm (Assume)}$$

$$N_{max} = 500 \text{ rpm (Assume)}$$

$$\frac{N_{max}}{N_{min}} = \phi^{n-1}$$

$$\frac{500}{160} = (\phi)^{6-1}$$

$$\phi = 1.256$$

We find $\phi = 1.256$ is not a standard ratio,

We can write, $1.12 \times 1.12 = 1.254$.

So $\phi = 1.12$ satisfies the requirement. Therefore the spindle speed term R 20 series, skipping one speed, are given by 160, 200, 250, 315, 400 and 500 rpm.

Structural Formula:

For 6 speeds, the preferred structural formula:

$$= 3(1) \times 2(3)$$

$$\frac{N_{min}}{N_{input}} = \frac{160}{250} \geq \frac{1}{4}$$

$$\frac{N_{max}}{N_{input}} = \frac{315}{250} \geq 2$$

Structural Formula:

$$= 2(1) \times 3(2)$$

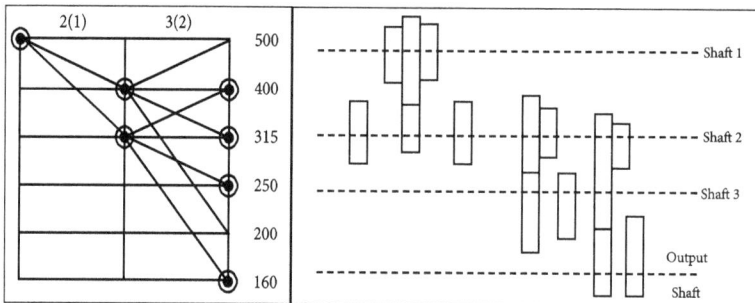

Conclusion:

Out of two schemes $3(1) \times 2(3)$ is better than other scheme. Because, only, $3(1) \times 2(3)$ scheme satisfies, the ratio requirement.

$$\frac{N_{max}}{N_{input}} < 2; \quad \frac{N_{min}}{N_{input}} < 1/4 \text{ and } \frac{N_{max}}{N_{min}} \leq 8$$

3. The spindle of a pillar drill is to run at 12 different speeds in the range of 100 rpm and 355 rpm. Design a three stage gear box with a standard step ratio. The gear box receives 5 kW from an electric motor running at 360 rpm. Sketch the layout of the gear box, indicating the number of teeth on each gear. Also, draw the sketch the speed diagram.

Given data:

$n = 12$

$$N_{min} = 100 \text{ rpm}$$

$$N_{max} = 355 \text{ rpm}$$

$$P = 51 \text{ kw}$$

$$N_{input} = 360 \text{ rpm}$$

$$\frac{N_{max}}{N_{min}} = (\phi)^{n-1}$$

$$\frac{355}{100} = (\phi)^{12-1}$$

Solution

$$\phi = 1.12$$

Spindle speed:

100, 112, 125, 140, 160, 180, 200, 224, 250, 280, 315, 355 rpm.

Structural Formula:

For 12 speeds:

$$3(1) \times 2(3) \times 2(6)$$

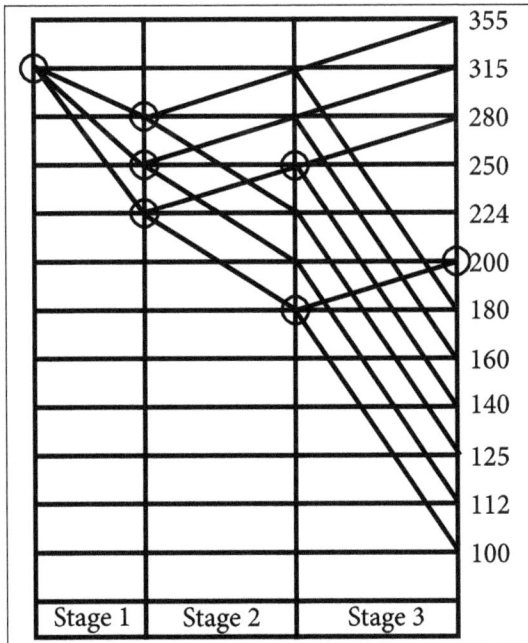

Ray Diagram.

Stage 3:

$$\frac{N_{min}}{N_{input}} = \frac{100}{180} = 0.5 > 1/4$$

$$\frac{N_{max}}{N_{input}} = \frac{200}{180} = 1.11 > 2$$

Stage 2:

$$\frac{N_{min}}{N_{input}} = \frac{180}{224} = 0.87 > 1/4$$

$$\frac{N_{max}}{N_{input}} = \frac{250}{224} = 1.116 < 2$$

Stage 1:

$$\frac{N_{min}}{N_{input}} = \frac{224}{315} = 0.74 < 1/4$$

$$\frac{N_{max}}{N_{input}} = \frac{280}{315} = 0.88 < 2$$

Number of Teeth

Stage 3:

Pair 1:

$$\frac{Z_{13}}{Z_{14}} = \frac{N_{14}}{N_{13}}$$

$$Z_{13} = Z_0 \ (\text{Assume})$$

$$\frac{Z_0}{Z_{14}} = \frac{100}{180}$$

$$Z_{14} = 36$$

Pair 2:

$$\frac{Z_{11}}{Z_{12}} = \frac{N_{12}}{N_{11}}$$

$$\frac{Z_{11}}{Z_{12}} = \frac{200}{100}$$

$$Z_{11} = 2 \times Z_{12}$$

$$Z_{11} + Z_{12} = Z_{13} + Z_{14}$$

$$2 Z_{12} + Z_{12} = 56.$$

$$3Z_{12} = 56$$

$$Z_{12} = 19 \,, \; Z_{11} = 38$$

Stage 2:

Pair 1:

$$\frac{Z_9}{Z_{10}} = \frac{N_{10}}{N_9}$$

$$Z_9 = Z_0 \text{ (Assume)}$$

$$\frac{Z_0}{Z_{10}} = \frac{180}{224}$$

$$Z_{10} = 25$$

$$\frac{Z_7}{Z_8} = \frac{N_8}{N_7}$$

$$\frac{Z_7}{Z_8} = \frac{250}{224}$$

$$\frac{Z_7}{Z_8} = 1.16$$

$$Z_7 = 1.16 \times Z_8$$

$$Z_7 + Z_8 = Z_9 + Z_{10}$$

$$1.16 \, Z_8 + Z_8 = 45$$

$$2.16 \, Z_8 = 45$$

$$Z_8 = 21, \; Z_7 = 24$$

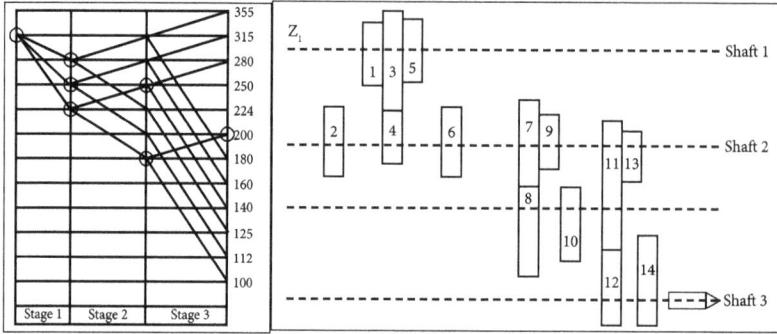

Stage 1:

First Pair:	Second Pair:	Third Pair:
$\dfrac{Z_5}{Z_6} = \dfrac{N_6}{N_5}$	$\dfrac{Z_1}{Z_2} = \dfrac{N_2}{N_1}$	$\dfrac{Z_3}{Z_4} = \dfrac{N_4}{N_3}$
$Z_o = 20\,(\text{Assume})$	$\dfrac{Z_1}{Z_2} = \dfrac{250}{315}$	$\dfrac{Z_3}{Z_4} = \dfrac{280}{315}$
$\dfrac{Z_0}{Z_6} = \dfrac{224}{315}$	$\dfrac{Z_1}{Z_2} = 0.79$	$Z_3 = 0.88 \times Z_4$
$Z_6 = 28$	$Z_1 = 0.79 \times Z_2$	$Z_3 = 0.88 \times 24$
$Z_5 = 20$	$Z_1 + Z_2 = 48$	$Z_3 + Z_4 = 48$
	$0.79\,Z_2 + Z_2 = 48$	$1.88 Z_4 = 48$
	$Z_2 = 27$	$Z_4 = 26$
	$Z_1 = 21$	$Z_3 = 23$

4.1.3 Kinematics layout

A kinematic layout is a pictorial representation of a gearbox, describing the arrangement of gears. It provides information like number of stages, number of shafts used number of gear pairs and its arrangement.

Problems

1. An all geared speed gear box is to be designed for a radial drilling machine with the following specifications:

Maximum size of the drill to be used = 50mm.

Minimum size of the drill to be used = 10mm.

Maximum cutting speed (drilling) = 40 m/min.

Minimum cutting speed (reaming, tapping, and boring) = 6m/min.

Number of speeds = 12.

Choose a 3 × 2 × 2 arrangement. Sketch the layout of the gearbox and the speed diagram. Calculate the percentage deviation of the obtainable speeds from the calculated ones.

Solution

$$n=12$$

$$N_{max} = 800 \text{ rpm}$$

$$N_{min} = 200 \text{ rpm}$$

Progression ratio:

$$\phi^{n-1} = \frac{N_{max}}{N_{min}}$$

$$\frac{800}{200} = \phi^{12.1}$$

$$\phi = 1.13 \text{ or}$$

$$\phi = 1.13$$

Permissible deviation $= \pm 10 \left(\phi - 1\right)\%.$

$$= \pm (1.13 - 1)\%$$

$$= 1.3$$

Action deviation $=(800-700)\times\dfrac{200}{800}$

$$= 22.5$$

If the actual deviation is more than the permission deviation, then non-standard speeds may be used.

The spindle speeds are: 200, 225.3, 254.5, 287.5, 325, 367, 415, 469, 530, 599, 676.87, 746 rpm.

Structure formula:

3(1) 3(3) 2(6)

Speed diagram:

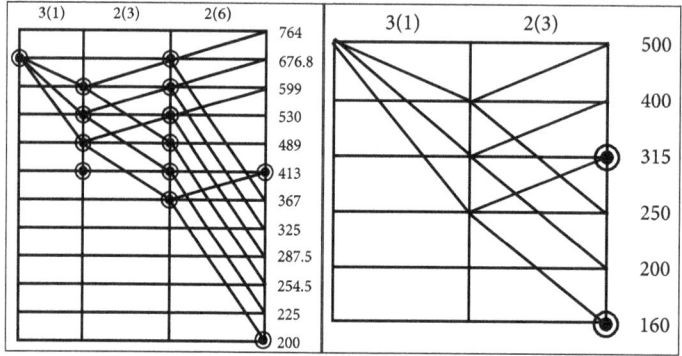

Structure formula = 3(1) 2(3) 2(6). Kinematic layout.

2. A design the layout of a 9 speed gear-box for a machine- tool. The minimum and maximum speeds are 10 and 90 rpm. Power is 5 kW from 1200 rpm. Induction motor. Let us draw the ray diagram and kinematic layout.

Given:

$$n = a \cdot N_{min} = 10_{rpm}; \ N = 90 \text{ rpm}$$

To find: Construction of speed diagram and kinematic layout.

Solution

Selection of spindle speed:

$$\frac{N_{max}}{N_{min}} = \phi^{n-1}$$

$$\frac{90}{10} = \phi^{9-1} \text{ or } \phi = (9)^{1/8} = 1.316$$

We can write, 1.316 is not a standard ratio.

So let us, find out multiples of standard ratio 1.12 or 1.06 comes closer to 1.316.

Then,

1.06×1.06×1.06×1.06×1.06 = 1.338 (skip 4 speeds).

$\phi = 1.06$. Satisfies the requirements.

Therefore, the spindle speeds from R40 series skipping 4 speeds given by:

10, 13, 18, 23, 31, 42, 56, 75, 100 rpm.

Structural Formula:

For 9 speed, the preferred structural formula = 3(1) 3(3).

Stage 2:

$$\frac{N_{min}}{N_{input}} = \frac{10}{31}$$

$$= 0.32 > \frac{1}{4}$$

$$\frac{N_{max}}{N_{input}} = \frac{56}{31}$$

$$= 1.8 < 2$$

∴ Ratio requirements satisfied.

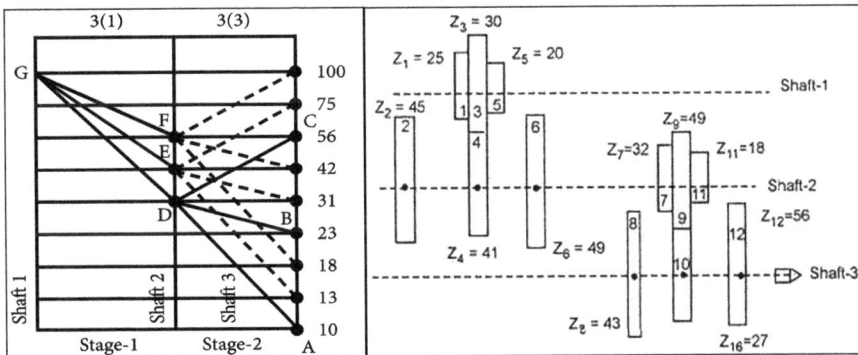

Speed diagram for 9 speed gear box Kinematic Arrangement.

Calculation of Number of Teeth

Let $Z_1, Z_2, Z_3, ..., Z_{12}$ = No. of teeth of gear 1, 2, 3, ... 12 respectively.

$N_1, N_2, N_3, ..., N_{12}$ = Speeds of the gear 1, 2, 3 ..., 12 respectively.

Second Stage

Pair 1:

First consider the ray that gives the maximum speed reduction. From the speed diagram, we find that the speed is reduced from 31 rpm to 10 rpm. We may assume that reduction is achieved by using the gears 11 and 12.

$$Z_{min} \geq 17 \text{ assume } Z_{11} = 18 \text{ (driver)}$$

$$\frac{Z_{11}}{Z_{12}} = \frac{N_{12}}{N_{11}} \text{ or } \frac{18}{Z_{12}} = \frac{10}{31}$$

$$Z_{12} = 55.8 \approx 56$$

Pair 2:

Minimum speed reduction from 31 rpm to 23 rpm. This can be achieved using gear T 7 and 8.

$$\frac{Z_7}{Z_8} = \frac{N_8}{N_7} = \frac{23}{31} Z_8 = 0.74 Z_8$$

Centre distance between the shafts are fixed and same. Therefore, the sum of number of teeth of moving gears should be equal.

$$Z_7 + Z_8 = Z_{11} + Z_{12}$$

$$0.74 Z_8 + Z_8 = 18 + 56$$

$$1.74 Z_8 = 74$$

$$Z_8 = 42.52 \approx 43$$

$$Z_7 = 31.8 \approx 32$$

Pair 3:

Speed increases from 31 rpm to 56 rpm.

$$\frac{Z_9}{Z_{10}} = \frac{N_{10}}{N_9} = \frac{56}{31}$$

$$Z_7 + Z_8 = Z_9 + Z_{10}$$

$$32 + 43 = 1.8 Z_{10} + Z_{10}$$

$$75 = 2.8 Z_{10}$$

$$Z_{10} = 26.7 \approx 27$$

$$Z_9 = 48.2 \approx 49$$

First Stage

Pair 1:

Consider maximum speed reduction from 75 rpm to 31 rpm. This can be achieved by gears 5 and 6.

$$\frac{Z_5}{Z_6} = \frac{N_6}{N_5} = \frac{31}{75}; \frac{20}{Z_6} = \frac{31}{75}; Z_5 = Z_0 \text{ (driver)}$$

$$Z_6 = 48.38 \approx 49$$

Pair 2:

Speed reduction from 75 rpm to 42 rpm.

$$\frac{Z_1}{Z_2} = \frac{N_2}{N_1} = \frac{42}{75} \text{ or } Z_1 = 0.56 \, Z_2$$

$$Z_1 + Z_2 = Z_5 + Z_6$$

$$0.56 \, Z_2 + Z_2 = 20 + 49$$

$$1.56 \, Z_2 = 69$$

$$Z_2 = 44.23 \approx 45$$

$$Z_1 = 24.76 \approx 25$$

Pair 3:

Speed reduction from 75 rpm to 56 rpm.

$$\frac{Z_3}{Z_4} = \frac{N_4}{N_3} = \frac{56}{75}$$

$$Z_3 = 0.74 \, Z_4$$

$$Z_1 + Z_2 = Z_3 + Z_4$$

$$25 + 45 = 0.74\, Z_4 + Z_4$$

$$70 = 1.74\, Z_4$$

$$Z_4 = 40.22 \approx 41$$

$$Z_3 = 29.77 \approx 30$$

A gear box is to be designed for the following specifications:

Power to be transmitted = 5.5 kW.

Number of speeds =9

Minimum speed = 280 rpm

Maximum speed = 1800 rpm

Input motor speed = 1400 rpm

3. Draw the kinematic layout diagram and the speed diagram. let us determine the number of teeth or all gears.

Given data:

       p = 5.5 kw

       n = 9 speeds

       $N_{min} = 280$ rpm

       $N_{max} = 1800$ rpm

       $N_{input} = 1400$ rpm

To find:

- Kinematic layout.
- Speed diagram.
- No. of teeth on each gear.

Solution

$$\text{Progression ratio}(\phi): \frac{N_{max}}{N_{min}} = (\phi)^{n+1}$$

$$\frac{N_{max}}{N_{min}} = \left(\phi\right)^{n-1}$$

$$\frac{1800}{280} = \left(\phi\right)^{a-1}$$

$\phi = 1.26$

$\phi = 1.26$ is not a standard step ratio

$= 1.12 \times 1.12$ (Skip one speed)

The spindle speeds are:

280, 355, 450, 560, 710, 900, 1120, 1400 and 1800 r.p.m

Structural Formula:

For a speed $= 3(1) \times 3(3)$

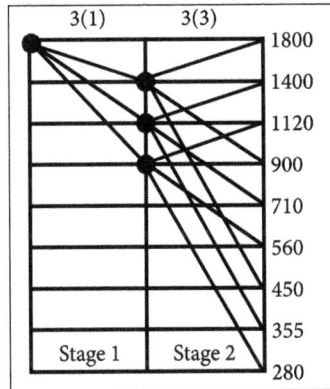

Ray Diagram.

Stage 1:

$$\frac{N_{min}}{N_{input}} = \frac{900}{1800} = 0.5 > 1/4$$

$$\frac{N_{max}}{N_{input}} = \frac{1400}{1800} = 0.77 < 2$$

Stage 2:

$$\frac{N_{min}}{N_{input}} = \frac{280}{900} = 0.311 > 1/4$$

$$\frac{N_{max}}{N_{input}} = \frac{1120}{900} = 1.24 < 2$$

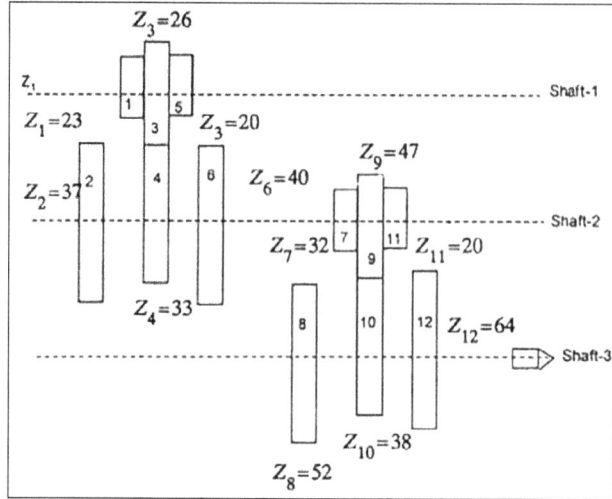

Kinematic Arrangement.

Calculation of Number of Teeth

Stage 2:

Pair 1:

$$\frac{Z_{11}}{Z_{12}} = \frac{N_{12}}{N_{11}}$$

$$Z_{11} = Z_0 \left(\text{Assume}\right)$$

$$\frac{Z_0}{Z_{12}} = \frac{280}{900}$$

$$Z_{12} = 64$$

Pair 2:

$$\frac{Z_7}{Z_8} = \frac{N_8}{N_7}$$

$$\frac{Z_7}{Z_8} = \frac{560}{900}; Z_7 = 0.02\ Z_8$$

$$Z_7 + Z_8 = Z_{11} + Z_{12}$$

$$0.62\ Z_8 + Z_8 = 84$$

$$1.62\ Z_8 = 84$$

$$Z_8 = \frac{84}{1.62}$$

$$Z_8 = 52$$

$$Z_7 = 32$$

Pair 3:

$$\frac{Z_9}{Z_{10}} = \frac{N_{10}}{N_9}$$

$$\frac{Z_9}{Z_{10}} = \frac{1120}{900}$$

$$\frac{Z_9}{Z_{10}} = 1.24$$

$$Z_9 = 1.24 \times Z_{10}$$

$$Z_9 + Z_{10} = Z_{11} + Z_{12}$$

$$1.24\,Z_{10} + Z_{10} = 84$$

$$2.24\,Z_{10} = 84$$

$$Z_{10} = \frac{84}{2.24}$$

$$Z_{10} = 38$$

$$Z_9 = 47$$

Stage 1:

Pair 1:

$$\frac{Z_3}{Z_6} = \frac{N_6}{N_5}$$

$$Z_5 = Z_0 \, (\text{Assume})$$

$$\frac{Z_0}{Z_6} = \frac{900}{1800}$$

$$Z_6 = 40$$

Pair 2:

$$\frac{Z_1}{Z_2} = \frac{N_2}{N_1}$$

$$\frac{Z_1}{Z_2} = \frac{1120}{1800}$$

$$Z_1 = 0.62 \times Z_2$$

$$Z_1 + Z_2 = Z_5 + Z_6$$

$$0.62 \times Z_2 + Z_2 = 60$$

$$1.62\, Z_2 = 60$$

$$Z_2 = 37$$

$$Z_1 = 23$$

Pair 3:

$$\frac{Z_3}{Z_4} = \frac{N_4}{N_3}$$

$$\frac{Z_3}{Z_4} = \frac{1400}{1800}$$

$$\frac{Z_3}{Z_4} = 0.77$$

$$23 = 0.77 \times Z_4$$

$$Z_3 + Z_4 = Z_5 + Z_6$$

$$0.77\, Z_4 + Z_4 = 60$$

$$Z_4 = 33$$

$$Z_3 = 26$$

4. A Select speeds for a 12 speed gear box for a minimum speed of 16 rpm and maximum speed of 900 rpm. Drive speed is 900 rpm. Draw speed diagram and draw kinematic arrangement of the gear box showing the number of teeth in all the gears.

Given:

n=12

$n_{min.} = 16$ rpm

$n_{max} = 900$ rpm

$n_{drive} = 900$ rpm

Solution

Step ratio, $\phi^{n-1} = \dfrac{n_{max}}{n_{min}}$

$$\phi = 1.44 \approx \dfrac{1.12 \times 1.12 \times 1.12}{\text{Skip 2 Speeds}}$$

The speeds are:

16, 22.4, 31.5, 45, 63, 90, 125, 180, 250, 355, 500, 710.

Structural Formula: $12 = 3(1) \cdot 2(3) \cdot 2(6)$

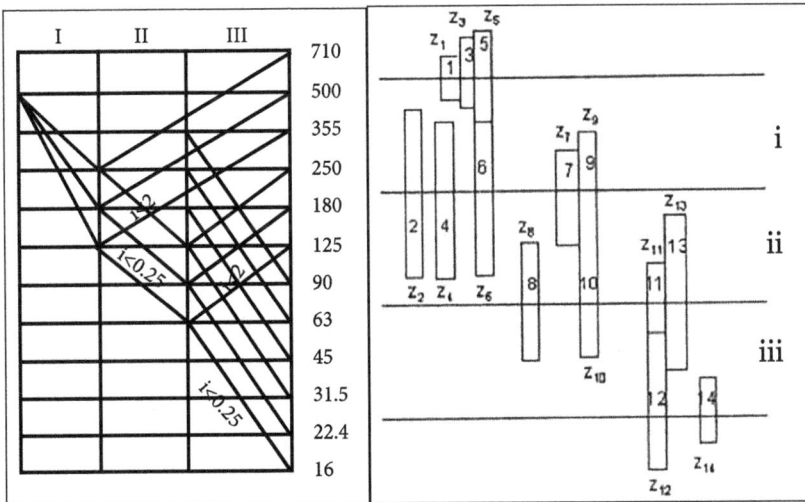

Ray Diagram and Kinematic Diagram.

Number of Teeth

Stage 1:

$$\frac{Z_1}{Z_2} = \frac{125}{710}; \frac{Z_3}{Z_4} = \frac{180}{710}; \frac{Z_5}{Z_6} = \frac{250}{710}$$

Assume:

$$Z_1 = Z_0$$
$$Z_2 = 114$$

$$Z_1 + Z_2 = Z_3 + Z_4 = Z_5 + Z_6 = 134$$

$$Z_4 = 107$$
$$Z_3 = 27$$

$$Z_6 = 99$$
$$Z_5 = 35$$

Stage 2:

$$\frac{Z_7}{Z_8} = \frac{63}{125}; \frac{Z_9}{Z_{10}} = \frac{180}{125}$$

Assume:

$$Z_7 + Z_8 = Z_9 + Z_{10} = 60$$

$$Z_7 = Z_0$$
$$Z_8 = 40$$

$$Z_{10} = 25$$
$$Z_9 = 35$$

Stage 3

$$\frac{Z_{11}}{Z_{12}} = \frac{16}{63}; \frac{Z_{13}}{Z_{14}} = \frac{125}{63}$$

Assume:

$$Z_{11} + Z_{12} = Z_{13} + Z_{14} = 99$$

$$Z_{11} = Z_0$$
$$Z_{12} = 79$$

$$Z_{14} = 33$$
$$Z_{13} = 66$$

4.2   Design of Sliding Mesh Gear Box

The clutch gear is rigidly fixed to the clutch shaft.

The clutch gear always remains connected to the drive gear of countershaft.

The other lay shaft gears are also rigidly fixed with it.

Two gears are mounted on the main shaft and can be sliding by shifter yoke when shifter is operated.

One gear is second & top speed gear and the other is the first and reverse speed gears. All gears used are spur gears.

A reverse idler gear is mounted on another shaft and always remains connected to reverse gear of counter shaft.

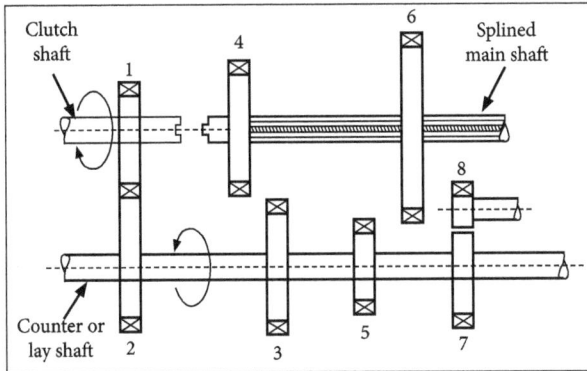

Sliding mesh gear box.

First Gear

- By operating gearshift lever, the larger gear on main shaft is made to slide and mesh with first gear of countershaft.

- The main shaft turns in the same direction as clutch shaft in the ratio of 3:1.

Second Gear

- By operating gear shaft lever, the smaller gear on the main shaft is made to slide and mesh with second gear of counter shaft.

- A gear reduction of approximately 2:1 is obtained.

Top Gear

- By operating gearshift lever, the combined second speed gear and top speed gear is forced axially against clutch shaft gear.

- External teeth on clutch gear mesh with internal teeth on top gear and the gear ratio is 1:1.

Reverse Gear

- By operating gearshift lever, the larger gear of main shaft is meshed with reverse idler gear.

- The reverse idler gear is always on the mesh with counter shaft reverse gear. Interposing the idler gear, between reverse and main shaft gear, the main shaft turns in a direction opposite to clutch shaft.

Neutral Gear

- When engine is running and the clutch is engaged, clutch shaft gear drives the drive gear of the lay shaft and thus lay shaft also rotates.

- But the main shaft remains stationary as no gears in main shaft are engaged with lay shaft gears.

Design Procedure

Step 1: Selection of standard pulley diameters. Calculate the diameters of the smaller and larger pulley using the relation: $i = \dfrac{D}{d} = \dfrac{n}{N}$ Then, select the standard pulley diameters from PSG 7.54.	D - Diameter of larger pulley (mm) N - Speed of the larger pulley (rpm) d - Diameter of small pulley (mm) n - Speed of the small pulley (rpm) i - velocity ratio.
Step 2: Calculation of design power. For calculating design power, select the load correction factor from PSG 7.53. The arc of contact factor from PSG 7.54 using the arc of contact value: Small diameter factor from PSG 7.62. Calculate the design power using the formula: $\text{Design Power} = \dfrac{Rated..Power \times Load..Correction..factor.}{Arc...of...contact..Factor \times Small...diameter...Factor.}$	Design Power - (kW) Rated. Power - Power of motor (kW).
Step 3: Selection of belt. Select the type of belt from PSG 7.52.	

Step 4: Load Rating and Number of plies. Load Rating Calculate the velocity of belt / belt speed using the formula: $$V = \frac{\pi dn}{60}.$$ Then, calculate the load rating using the formula in PSG 7.54. Number of plies, select the number of plies required from PSG 7.52.	d - Diameter of small pulley (mm) n - Speed of the small pulley (rpm) V-Velocity of belt of speed of belt (m/s) Load rating - (kW/mm/ply).
Step 5: Belt Width. Calculate the belt width using the formula: $$Width...of...pulley = \frac{Design...power}{Load...rating...\times...No...of...plies}$$ Select the standard belt width from PSG 7.52.	Design Power - (kW) Load rating - (kW/mm/ply) Width of belt - (mm).

Problems

1. Let us the design a sliding mesh nine speed gear box for a machine tool with speed ranging from 36 rpm to 550 rpm. Draw the speed diagram and kinematic arrangement showing number of teeth in all gears.

Given:

$$n = 9$$

$$n_{min} = 36 \text{ rpm}$$

$$n_{max} = 550 \text{ rpm}$$

$$\text{Step ratio, } \phi^{n-1} = \frac{n_{max}}{n_{min}}$$

$$\phi = 1.406 \approx \frac{\left|1.12 \times 1.12 \times 1.12\right|}{\text{Skip 2 speeds}}$$

Solution

The speeds are:

35.5, 50, 71, 100, 140, 200, 280, 400, 560.

Structural Formula: $9 = 3(1) \cdot 3(3)$.

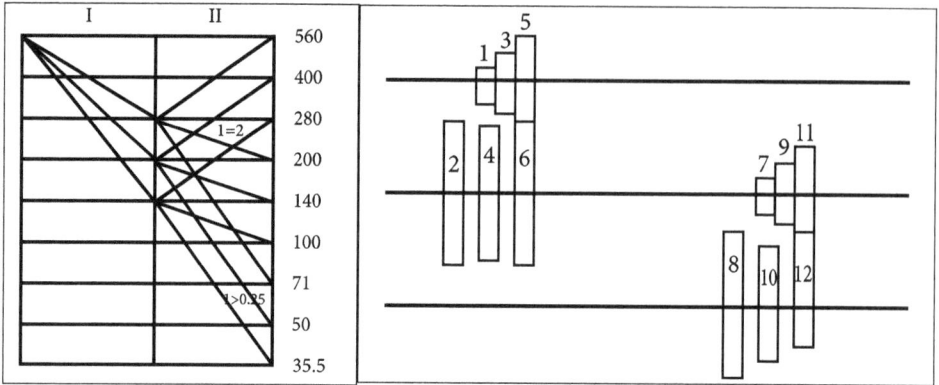

Ray Diagram and Kinematic Arrangement.

Number of Teeth

Stage 1:

$$\frac{Z_1}{Z_2} = \frac{140}{560}; \frac{Z_3}{Z_4} = \frac{200}{560}; \frac{Z_5}{Z_6} = \frac{280}{560}$$

Take $Z_1 + Z_2 = Z_3 + Z_4 = Z_5 + Z_6 = 100$

$Z_1 = Z_0$

$Z_2 = 80$

$Z_3 = 26$

$Z_4 = 74$

$Z_5 = 33$

$Z_6 = 67$

Stage 2

$$\frac{Z_7}{Z_8} = \frac{33.5}{140}; \frac{Z_9}{Z_{10}} = \frac{100}{140}; \frac{Z_{11}}{Z_{12}} = \frac{280}{140}$$

Take, $Z_7 + Z_8 = Z_9 + Z_{10} = Z_{11} + Z_{12} = 99$

$Z_7 = Z_0$

$Z_8 = 79$

$Z_9 = 41$

$Z_{10} = 58$

$Z_{11} = 66$

$Z_{12} = 33$

4.3 Design of Multi Speed Gear Box for Machine Tool Applications

If gear box operates with more than one speed ratio or gear ratio, it is called as multi-speed gearbox. The machine tool need to operate with different spindle speeds. However, the prime mover, which is an electric-motor, operates at a single constant speed. Therefore, for obtaining number of output speeds with a single input speed, the multi-speed gearbox is required in machine tools.

The used for speed adjustment at constant power level, this range of Heavy duty Industrial Multi-speed gearboxes are designed for continuous operation.

Based on standard cast-iron castings, the versatile design allows to offer several options:

- The output shafts can be supplied to rotate in the same or opposite directions.

- Motor flanges can be supplied for direct motor mounting.

- Up to 288 different speeds when using 4 gearboxes coupled together.

Possible Arrangements to Achieve 12 Speeds from a Gear Box

3(1) 2(3) 2(6).

(OR)

2(1) 3(2) 2(6).

(OR)

2(1) 2(2) 3(4).

Application

Turner Uni-drive industrial gearboxes and multi-speed transmissions are used throughout the world for a wide variety of applications ranging from wire drawing in New Zealand and tube processing in Oman to food packaging in Pennsylvania and petroleum processing in Wyoming.

Torque multiplication through a mechanical gearbox will typically generate long term energy savings and lower first costs.

Some of the applications of gear box are:

- Steel and tube processing.

- Food processing.

- Wire drawing.

- Plastic extrusion.

- Aggregate and construction.

- Petroleum processing.

- Test stands.

- Waste water treatment.

Problems

1. The design a 12 speed gear box for an all geared head stock of a lathe. Maximum and minimum speeds are 600 rpm and 25 rpm respectively. The drive is from an electric motor giving 2.25 kw at 1440 rpm.

Given data:

> 12 = Speed gear box

> (N_{min}) Maximum Speed = 25 rpm

> (N_{max}) Maximum Speed = 600 rpm

> Power = 2.25 kW

> N = 1440 rpm

Find out:

(1) Kinematic Diagram.

(2) Ray Diagram, etc.

$$\phi = \text{Speed ratio} = \left(\frac{N_{max}}{N_{min}}\right)^{\frac{1}{Z-1}} = \left(\frac{600}{25}\right)^{\frac{1}{12-1}} = 1.33$$

Solution

> $\phi = 1.33$

> $N_1 = 25$ rpm; $N_2 = 31.5$ rpm; $N_3 = 40$ rpm; $N_4 = 50$ rpm

> $N_5 = 63$ rpm; $N_6 = 80$ rpm; $N_7 = 100$ rpm; $N_8 = 125$ rpm

> $N_9 = 160$ rpm; $N_{10} = 200$ rpm; $N_{11} = 250$ rpm; $N_{12} = 315$ rpm

Structural Formula for 12 Speed is given by,

$$P_1(X_1) \; P_2(X_2) \; P_3(X_3)$$

3(1)2(3)2(6)

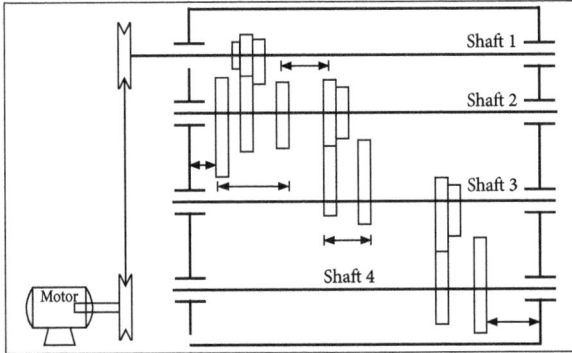

Kinematic Arrangement Diagram 3(1) 2(3) 2(6).

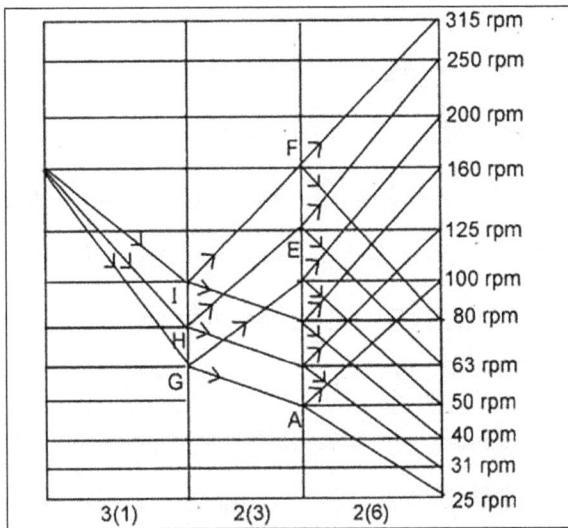

Ray Diagram - 3(1) 2(3) 2(6).

To Select, Input Speed for the last (3rd) Stage:

$$\frac{N_{max}}{N_{input}} \le 2 \; ; \frac{N_{min}}{N_{input}} \ge 0.25$$

Consider, 50 rpm as input speed:

$$\frac{N_{max}}{N_{input}} = \frac{80}{50} = 1.6 < 2 \text{ design is safe.}$$

$$\frac{N_{min}}{N_{input}} = \frac{25}{50} = 0.5 < 0.25 \text{ design is safe.}$$

So, select 50 rpm as the input speed for the third stage shaft.

For Second Stage, Select Input Speed as mentioned below:

$$\frac{N_{max}}{N_{input}} \le 2; \frac{N_{min}}{N_{input}} \le 0.25$$

Select 63 rpm as input speed for the second stage:

$$\frac{N_{max}}{N_{input}} = \frac{100}{63} = 1.5 < 2 \text{ Selection is safe.}$$

$$\frac{N_{min}}{N_{input}} = \frac{50}{63} = 0.79 > 0.25 \text{ Selection is safe.}$$

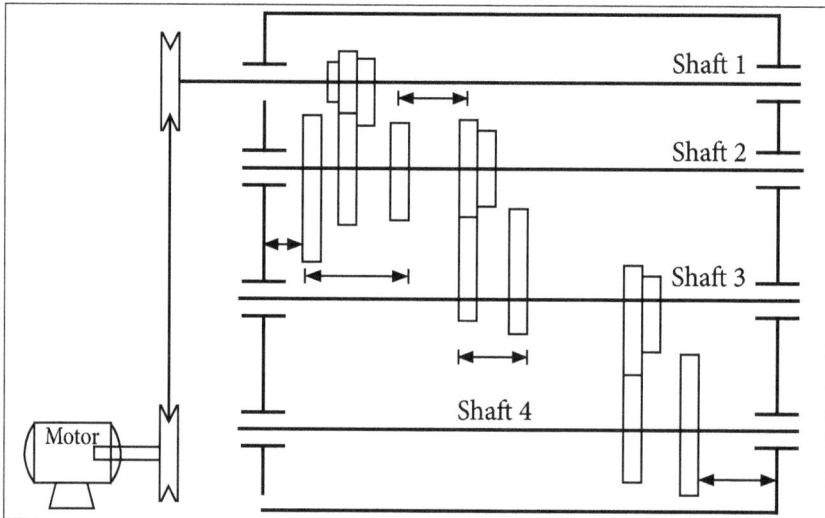

Kinematic Layout of 12 Speed Gear Box.

As mentioned in the previous problem all the stage gear's number of teethes are calculated.

2. Let us in a machine tool application, 12 different speeds are required from 125 rpm to 450 rpm in the output shaft. The motor speed is 630 rpm.

- Determine the 12 standard speed in G.P.

- Draw the ray diagram and kinematic layout.

- Determine the number of teeth on the gears to be used.

Given:

$$N = 12; N_{min} = 125 \text{ rpm}; N_{max} = 450 \text{ rpm}$$

$$N_{input} = 630 \text{ rpm}$$

(1) Determination of 12 spindle speeds:

$$\frac{N_{max}}{N_{min}} = \phi^{n-1}$$

$$\frac{450}{125} = \phi^{12-1} \text{ or } \phi = 1.123$$

Since the calculated ϕ (= 1.123) is a standard step ratio for R20 series, therefore spindle speed.

125, 140, 160, 180, 200, 224, 250, 280, 315, 355, 400 and 450 rpm.

(2) Ray Diagram:

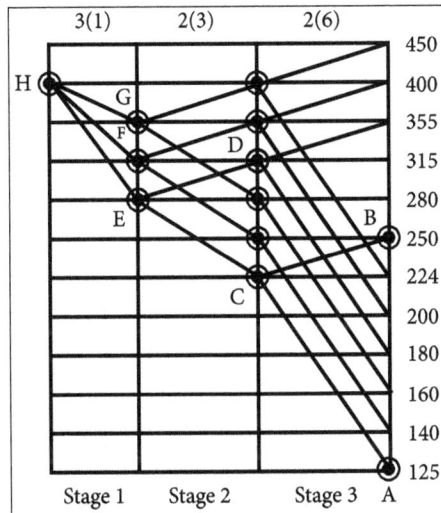

Ray Diagram for 12 speed gear box.

Stage 3:

$$\frac{N_{min}}{N_{input}} = \frac{125}{200}$$

$$= 0.625 > \frac{1}{4}$$

$$\frac{N_{max}}{N_{input}} = \frac{250}{200}$$

$$= 1.25 < 2$$

Stage 2:

$$\frac{N_{min}}{N_{input}} = \frac{200}{250} = 0.8 > \frac{1}{4}$$

$$\frac{N_{max}}{N_{input}} = \frac{280}{250} = 1.12 > 2$$

Stage 1:

$$\frac{N_{min}}{N_{input}} = \frac{250}{400} = 0.625 < \frac{1}{4}$$

$$\frac{N_{max}}{N_{input}} = \frac{315}{400} = 0.787 < 2$$

$\therefore$ Ratio requirements are satisfied.

(3) Kinematic Arrangement:

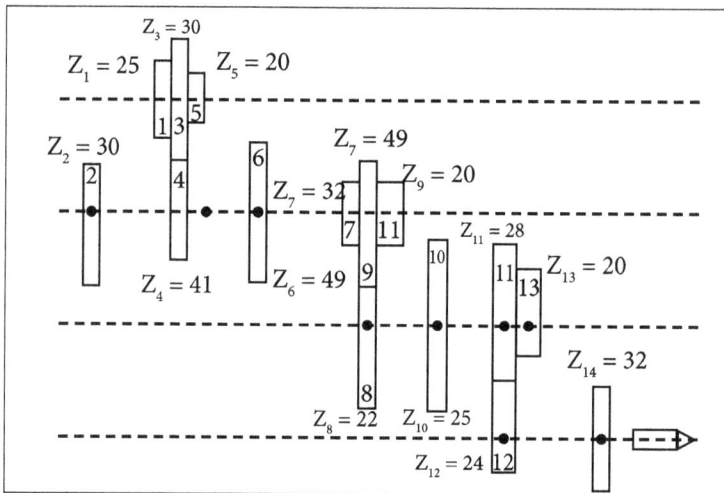

Kinematic Arrangement for 12 Speed Gear Box.

(4) Number of Teeth on the Gears to be Used:

Stage 3:

1st pair:

Speed reduces from 200 rpm to 125 rpm gears 13 and 14.

$$Z_{min} \geq 17 \quad Z_{13} = Z_0 \, (\text{driven}), \text{Assumed}$$

$$\frac{Z_{13}}{Z_{14}} = \frac{N_{14}}{N_{13}} ; \frac{Z_0}{Z_{14}} = \frac{125}{200}$$

$$\therefore Z_{14} = 32$$

2nd pair:

$$\frac{Z_{11}}{Z_{12}} = \frac{N_{12}}{N_{11}} = \frac{250}{200} \text{ or } Z_{11} = 1.25 \, Z_{12}$$

We know that,

$$Z_{11} + Z_{12} = Z_{13} + Z_{14} = 20 + 32 = 52 \qquad \qquad ...(i)$$

From (i) and (ii),

$$Z_{12} = 24 \; ; \; Z_{11} = 28 \qquad \qquad ...(ii)$$

Stage 2:

Pair 1:

Assume $Z_9 = Z_0 \, (\text{driver})$

$$\frac{Z_9}{Z_{10}} = \frac{N_{10}}{N_9} ; \frac{Z_0}{Z_{10}} = \frac{200}{250}$$

$$\therefore Z_{10} = 25$$

Pair 2:

$$\frac{Z_7}{Z_8} = \frac{N_8}{N_7} = \frac{250}{250} \text{ or } Z_7 = 1.12 \, Z_8 \qquad \qquad ...(iii)$$

$$Z_7 + Z_8 = Z_9 + Z_{10} = 20 + 25 = 45 \qquad \qquad ...(iv)$$

From (iii) and (iv), $Z_8 = 22 \; ; \; Z_7 = 23$.

Stage 1:

Pair 1: Assume $Z_5 = Z_0 \, (\text{driver})$

$$\frac{Z_5}{Z_6} = \frac{N_6}{N_5} ; \frac{Z_0}{Z_6} = \frac{250}{400}$$

$$\therefore Z_6 = 32$$

Pair 2:

$$\frac{Z_3}{Z_4} = \frac{N_4}{N_3} = \frac{280}{400} \text{ or } Z_3 = 0.74 \, Z_4$$

$$Z_2 + Z_4 = Z_5 + Z_6 = 20 + 32 = 52$$

$$Z_4 = 31 \text{ and } Z_3 = 21$$

Pair 3:

$$\frac{Z_1}{Z_2} = \frac{N_2}{N_1} = \frac{315}{400} \text{ or } Z_1 = 0.787\ Z_2$$

$$Z_1 + Z_2 = Z_3 + Z_4 = 52$$

$$Z_2 = 30\ ;\ Z_1 = 22$$

4.4   Constant Mesh Gear Box

The constant mesh gearbox is a type of transmission in which all or most of the gears are always in mesh with one another as opposed to a sliding-gear transmission, in which engagement is obtained by sliding some of the gears along a shaft into mesh.

In a constant-mesh manual gearbox, Gear ratios are selected by small Clutches that connect the various gear sets to their shafts so that power is transmitted through them.

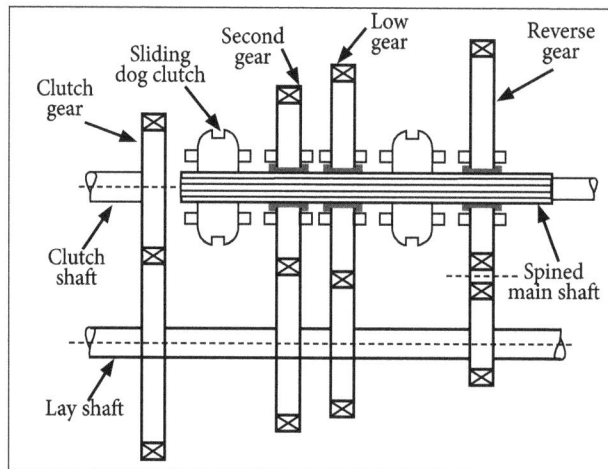

In this type of gearbox, all the gears of the main shaft are in constant mesh with corresponding gears of the countershaft.

*   The gears on the main shaft which are bushed are free to rotate.

*   The dog clutches are provided on main shaft.

*   The gears on the lay shaft are, however, fixed.

- When the left Dog clutch is slide to the left by means of the selector mechanism, its teeth are engaged with those on the clutch gear and we get the direct gear.

- The same dog clutch, however, when slide to right makes contact with the second gear and second gear is obtained.

- Similarly movement of the right dog clutch to the left results in low gear and towards right in reverse gear. Usually the helical gears are used in constant mesh gearbox for smooth and noiseless operation.

Problems

1. A load lifting arrangement transmitting 10 kW with electric motor running at 1400 rpm, constant mesh type speed reducer is required with reduction ratio 12. Let us Design a suitable arrangement and make a neat sketch. (All value refer to PSG data book).

Given:

$P = 10 \text{ kW}$

$N_1 = 1400 \text{ rpm}$

$i = 12$

Solution

Step 1:

Assume Work-Steel.

Worm Wheel - Bronze.

Assume, $V = 3 \text{ m/s}$.

Step 2:

Center distance,

$$a = \left(\frac{Z}{q}+1\right)3\sqrt{\left[\dfrac{540}{\dfrac{Z}{q}[\sigma_c]}\right]^2 [M_t]}.$$

Assume, $Z = 2$.

$$i = \frac{Z}{2} \Rightarrow z = 24$$

Initially, choose q = 11.

$$[\sigma_c] = 1590 \text{ kgf/cm}^2$$

$$[M_t] = 1 \times M_t = 1 \times \frac{P \times 60}{2\pi n_2}$$

$$i = \frac{n_1}{n_2} \Rightarrow n_2 = 116.67 \text{ rpm}$$

$$[M_t] = 8184.8 \text{ kgf cm}$$

a = 18.55

a = 20 cm

Step 3:

Axial module,

$$M_x = 1.24\ ^3\!\sqrt{\frac{[M_t]}{Z\ q\ y_v\ [\sigma_b]}}$$

$$Z_V = \frac{Z}{\cos^3 \gamma}$$

$$\gamma = \tan^{-1}\left(\frac{Z}{q}\right)$$

$$= 10.3^\circ$$

$$\Rightarrow Z_V = 42$$

$$y_v = 0.471$$

$$[\sigma_b] = 780 \text{ kgf/cm}^2$$

$$m_x = 0.544$$

$$m_x = 0.6 \text{ cm}$$

Step 4:

$$\sigma_c = \frac{540}{\left(\dfrac{Z}{q}\right)} \sqrt{\left[\frac{\dfrac{Z}{q}+1}{a}\right]^3 [M_t]}$$

$$\sigma_c = 1420.8 \text{ kgf/cm}^2 < \left[\sigma_c\right]$$

$$\sigma_b = \frac{1.9[M_t]}{m_x^3 \, q \, Z \, y_v}$$

$$\sigma_b = 579 \text{ kgf/cm}^2 < \left[\sigma_b\right]$$

Step 5:

Face width of wheel,

b = 0.7 dx

= 0.7 (q mx)

= 4.62 cm

Length of Worm, L ≥ (11 + 0.6 Z) mx

L ≥ 7.464 cm

L = 8 cm.

4.5   Speed Reducer Unit

Methods of Lubrication in Speed Reducer:

- Oil bath by dipping (or) splashing.

- Oil pressure fed by using gear pump.

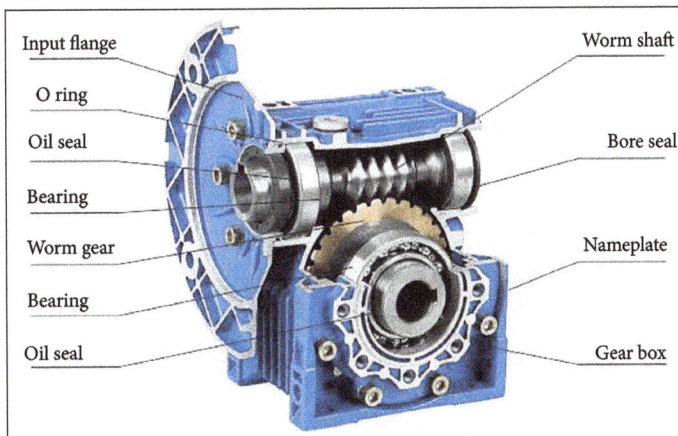

The gear speed reducers are comprised of the terms "gearbox" and "speed reducer" that are used interchangeably in the world of power transmission and motion control. Gearboxes are used for speed reduction and torque multiplication. The term speed reducer became vernacular when gearboxes were first implemented in industry.

The speed reduction was an important function for the gearbox, to replace more cumbersome belts and pulleys technology. Demand for worm gear speed reducers is increasing as more mechanical applications in several industries require speed reduction, ranging from rock crushers to robots.

4.5.1 Variable Speed Gear Box

There are many and diverse reasons for using variable speed drives. Some applications, such as paper making machines, cannot run without them while others such as centrifugal pumps, can benefit from energy savings.

In general, variable speed drives are used to:

- Latch the speed of a drive to the process requirements.

- Latch the torque of a drive to the process requirements.

- Save energy and improve efficiency.

4.6   Fluid Couplings

A fluid coupling or hydraulic coupling is a hydrodynamic device used to transmit rotating mechanical power. It has been used in automobile transmissions as an alternative to a mechanical clutch. It also has widespread application in marine and industrial machine drives, where variable speed operation and controlled start-up without shock loading of the power transmission system is essential.

Operating Principle

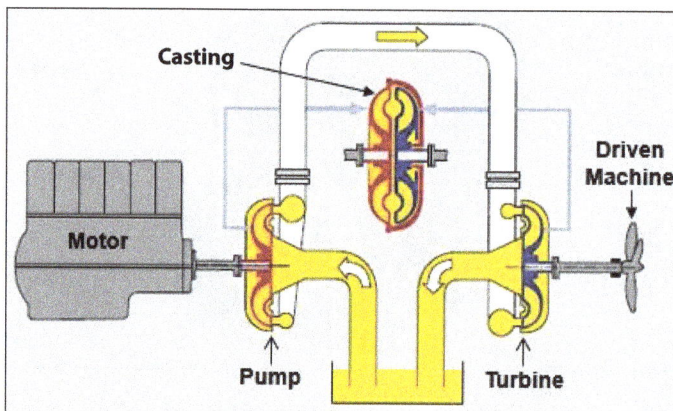

There is no mechanical interconnection between the impeller and the rotor (i.e. the driving and driven units) and the power is transmitted by virtue of the fluid filled in the coupling. The impeller when rotated by the prime mover imparts velocity and energy to the fluid, which is converted into mechanical energy in the rotor thus rotating it.

The fluid follows a closed circuit of flow from impeller to rotor through the air gap at the outer periphery and from rotor to impeller again through the air gap at the inner periphery. To enable the fluid to flow from impeller to rotor it is essential that there is difference in the "head" between the two and thus it is essential that there is difference in R.P.M. known as slip between the two.

Slip is an important and inherent characteristic of a fluid coupling resulting in several desired advantages. As the slip increases more and more fluid can be transferred from the impeller to the rotor and more torque is transmitted. However when the rotor is at standstill, maximum fluid is transmitted from the coupling. The maximum torque is limiting torque. The fluid coupling also acts as a torque limiter.

4.7   Torque Converters for Automotive Applications

The manual transmissions, we know that an engine is connected to a transmission by way of a clutch. Without this connection a car would not be able to come to a complete stop without killing the engine. But cars with an automatic transmission have no clutch that disconnects the transmission from the engine. Instead, they use an amazing device called a torque converter.

It may not look like much, but there are some very interesting things going on inside.

Basics

The automatic transmissions need a way to let the engine turn while the wheels and gears in the transmission come to a stop. A manual transmission cars use a clutch, which completely disconnects the engine from the transmission. Automatic transmission cars use a torque converter.

A torque converter is a type of fluid coupling, which allows the engine to spin somewhat independently of the transmission. If the engine is turning slowly, such as when the car is idling at a stoplight, some amount of torque passed through the torque converter is very small, so keeping the car still requires only a light pressure on the brake pedal.

If we were to step on the gas pedal while the car is stopped, you would have to press harder on the brake to keep the car from moving. This is because when you step on the gas, the engine speeds up and pumps more fluid into the torque converter, causing more torque to be transmitted to the wheels.

Inside a Torque Converter

As shown in the figure given below, there are 4 components inside the very strong housing of the torque converter:

- Pump

- Turbine

- Stator

- Transmission fluid.

The housing of the torque converter is bolted to the flywheel of the engine so it turns at whatever speed the engine is running at. The fins that make up the pump of the torque converter are attached to the housing, so they also turn at the same speed as the engine. The cutaway below shows how everything is connected inside the torque converter.

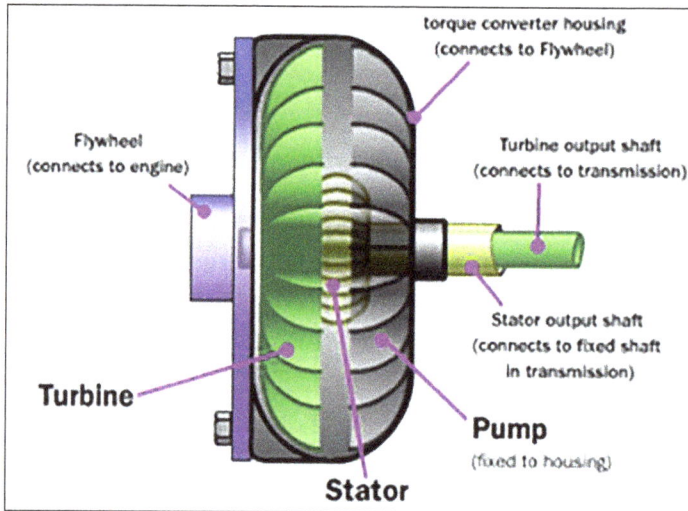

The pump inside a torque converter is a type of centrifugal pump. As it spins, fluid is flung to the outside, much as the spin cycle of a washing machine flings water and clothes to the outside of the wash tub. As fluid is flung to the outside, a vacuum is created that draws more fluid in at the center. The fluid then enters the blades of the turbine, which is connected to the transmission.

The turbine causes the transmission to spin, which basically moves your car. We can see in the graphic below that the blades of the turbine are curved. The means that the fluid, which enters the turbine from the outside, has to change direction before it exits the center of the turbine. It is this directional change that causes the turbine to spin. In order to change the direction of a moving object, we must apply a force to that object it doesn't matter if the object is a car or a drop of fluid. And whatever applies the force that causes the object to turn must also feel that force, but in the opposite direction.

So as the turbine causes the fluid to change direction, the fluid causes the turbine to spin. The fluid exits the turbine at the center, moving in a different direction than when it entered. If we look at the arrows in the figure above, we can see that the fluid exits the turbine moving opposite the direction that the pump (and engine) are turning. If the fluid were allowed to hit the pump, it would slow the engine down, wasting power.

Cams, Clutches and Brakes

5.1   CAM Design

The transformation of one of the simple motions, such as rotation into any other motions is often conveniently accomplished by means of a cam mechanism A cam mechanism usually consists of two moving elements, the cam and the follower, mounted on a fixed frame. Cam devices are versatile, and almost any arbitrarily-specified motion can be obtained. In some instances, they offer the simplest and most compact way to transform motions.

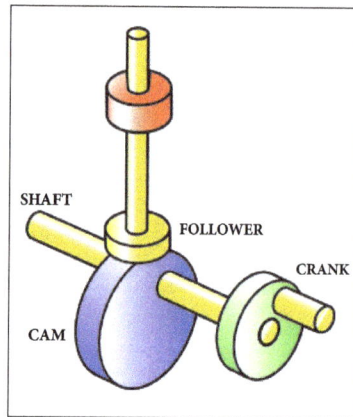

A cam may be defined as a machine element having a curved outline or a curved groove, which by its oscillation or rotation motion, gives a predetermined specified motion to another element called the follower.

The cam has a very important function in the operation of many classes of machines, especially those of the automatic type, such as printing presses, shoe machinery, textile machinery, gear cutting machines and screw machines. In any class of machinery in which automatic control and accurate timing are paramount. The cam is an indispensable part of the mechanism. The possible applications of cams are unlimited and their shapes occur in great variety.

Base Circle

It is the smallest circle drawn to the cam positive from the center of station of a radial cam The size of the cam drive depends on the size of the base circle.

Pitch Circle

It is a circle with its center as the center of cam axis and radius such that it passes through the pitch point.

Problems

1. A draw the displace time, Velocity time and the acceleration time curves for the follower in order to satisfy the following conditions:

- Stroke of the follower 25 mm.

- Out stroke takes place with SHM during 90° of cam rotation.

- Return stroke takes with SHM during 75° of cam rotation.

- Cam rotates with a uniform speed of 800 rpm.

Solution

Given:

Stroke of the follower, S = 25 mm. = 0.025 m

$$\text{Outstroke } \theta_o = 90° \frac{90 \times \pi}{180} = 1.51 \text{ rad/s.}$$

$$\text{Return stroke } \theta_R = 75°. \frac{75 \times \pi}{180} = 1.31 \text{ rad/s}$$

Speed N = 800 rpm

Time required for the outstroke of the follower in seconds,

$$t_o = \frac{\theta_o}{\omega} = \frac{1.571}{83.77} = 0.018 \text{ m.}$$

$$(\omega) = \frac{2\pi N}{60} = \frac{2 \times \pi \times 800}{60}$$

$$= 83.77 \text{ m/s}$$

Maximum velocity of the follower on the out stroke:

$$V_o = \frac{\pi \omega S}{2\theta_o} = \frac{\pi \times 83.77 \times 0.025}{2 \times 1.571} = 2.09 \text{ m/s}$$

Maximum acceleration of the follower on the outstroke:

$$a_o = \frac{V_o^2}{OP} = \frac{\pi^2 \omega^2 S}{2(\theta_o)^2} = \frac{\pi^2 \times 83.77^2 \times 0.025}{2(1.571)^2} = 350.7 \text{ m/s.}$$

Maximum acceleration of the follower return stroke:

$$V_R = \frac{\pi \omega S}{2\theta_R} = \frac{\pi \times 83.77 \times 0.025}{2 \times 1.31} = 2.5 \text{ m/s}$$

Maximum acceleration of the follower return stroke:

$$a_R = \frac{\pi^2 \omega^2 S}{2(\theta_R)^2} = \frac{\pi^2 \times 83.77^2 \times 0.025}{2(1.31)^2} = 504.4 \text{ m/s}^2$$

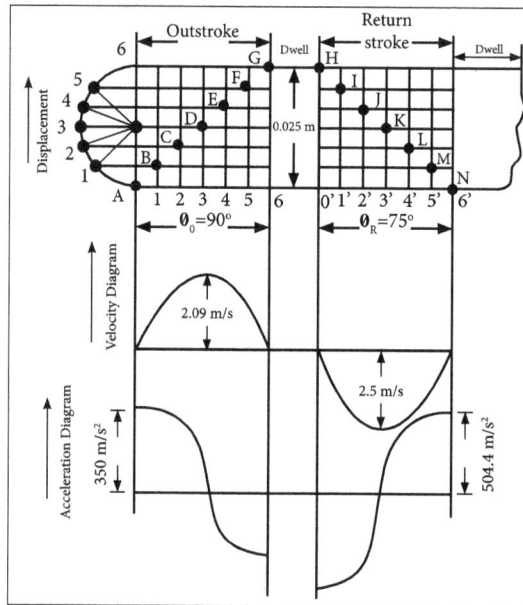

2. A cam is to give the following motion to a knife-edged follower:

- Out stroke during 60° of cam rotation.

- Dwell for the next 30° of cam rotation.

- Return stroke during next 60° of cam rotation.

- Dwell or the remaining 210° of cam rotation. The stroke of the follower is 40 mm and the minimum radius of the cam is 50 mm; The follower moves with uniform velocity during both the out stroke and return strokes. Draw the profile of the cam when the axis of the follower passes through the axis of the cam shaft.

Given Data:

Knife edge follower.

Uniform velocity.

$$\theta_R = 60°$$

$$\theta_D = 30°$$

$$\theta_F = 60°$$

Solution

Scale

X-axis, 1 cm = 20°.

Y-axis, 1 cm = Actual scale.

Displacement Diagram.

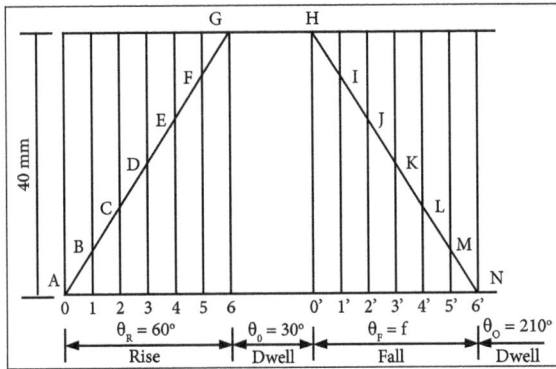

$$\theta_D = 210°$$

Min. Radius = 50 mm

Stroke of the follows L = 40 mm

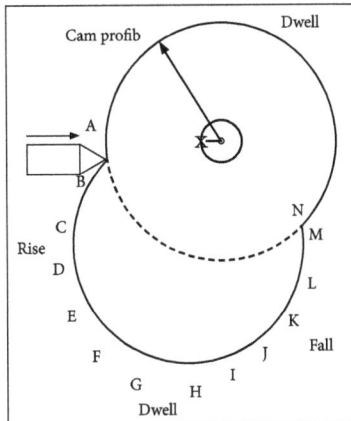

3. A radical cam rotating at 200 rpm is driving a 10 mm diameter translating roller follower to produce the following motions, rise of 20 mm with SHM in 150° of cam

rotation, dwell for 66° and fall of 20 mm with SHM in 120° of cam rotation and dwell for remaining 30°. Draw the profile of the cam. Check whether under cutting will occur.

Given:

Roller Follower,

N = 200 RPM ; Roller Dia = 10mm,

Rise = Fall [Of Roller] = S = 20 mm,

Forward Stroke $\left[\partial_F \right] = 150°$; Simple Harmonic Motion [SHM].

1st Dwell Angle = 60°;

Return Stroke Angle $\left[\partial_R \right] = 120°$; [SHM].

2nd Dwell Angle = 30°.

Solution

Undercutting:

$$e_{k_{min}} = \frac{\left[(R_P + y)^2 + \left(\frac{1}{W} \frac{dy}{dt} \right)^2 \right]^{3/2}}{(R_P + y)^2 + 2 \left(\frac{1}{W} \frac{dy}{dt} \right)^2 - (R_P - y) \left(\frac{1}{W^2} \frac{d^2 y}{dt^2} \right)}$$

$$w = \frac{2\pi N}{60} = 20.94 \text{ rad/S} \qquad\qquad …(1)$$

Maximum negative acceleration occurs at $\partial / \beta = 1$.

$$y = \frac{h}{2} \left(1 - \cos \frac{\pi \theta}{\beta} \right) = 20 \text{ mm}$$

$$\frac{dy}{dt} = \frac{h \pi W}{2\beta} \left(\sin \frac{\pi \theta}{\beta} \right) = 0$$

$$\frac{d^2 y}{dt^2} = \frac{h}{2} \left(\frac{\pi W}{\beta} \right)^2 \cos \frac{\pi \theta}{\beta}$$

$$= -6314.16$$

Assuming base coils radius,

$$R_b = 20 \text{ ram}$$

$$R_P = R_b + R_\gamma = 25 \text{ mm}$$

$$\Rightarrow \rho k_{min} = 34mm > R_r = 5 \text{ mm}$$

There is no underwriting.

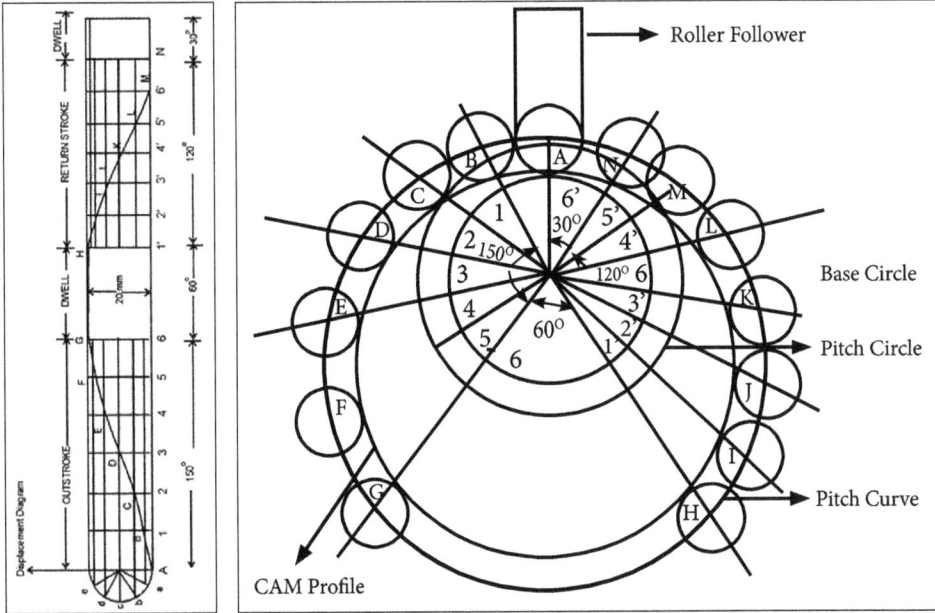

CAM Profile

4. Design a cam for operating the exhaust valve of an oil engine. It is required to give equal uniform acceleration and retardation during opening and closing of the valve, each of which corresponding to 60° of cam rotation. The valve should remain in the fully Open position for 20° of cam rotation. The lift of the valve is 32 mm and the least radius of the cam is 50 mm, the following is provided with a roller of 30 mm diameter and its line of stroke passes through the axis of the cam.

Displacement Diagram.

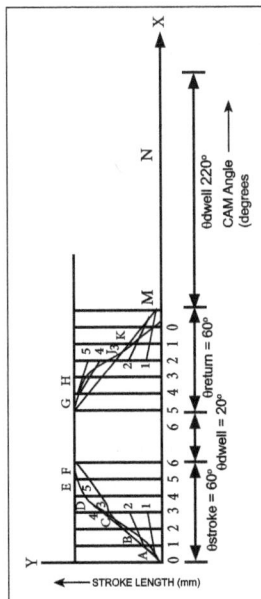

Cam Profile

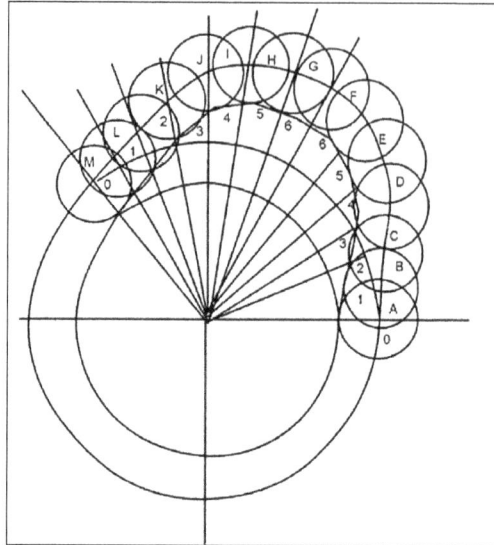

5.1.1 Types of CAM

Cams can be classified based on their physical shape.

(a) Disk or Plate CAM

The disk (or plate) cam has an irregular contour to impart a specific motion to the follower. The follower moves in a plane perpendicular to the axis of rotation of the camshaft and is held in contact with the cam by springs or gravity.

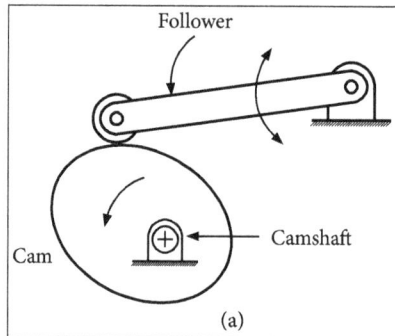

(b) Cylindrical CAM

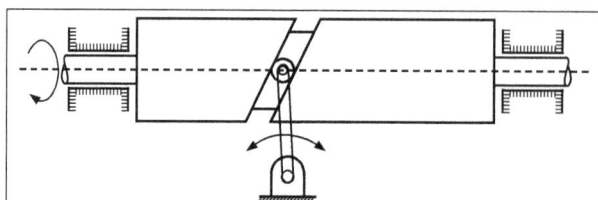

The cylindrical cam has a groove cut along its cylindrical surface. The roller follows the groove, and the follower moves in a plane parallel to the axis of rotation of the cylinder.

(c) Translating CAM

The translating cam is a contoured or grooved plate sliding on a guiding surface(s). The follower may oscillate or reciprocate. The contour or the shape of the groove is determined by the specified motion of the follow:

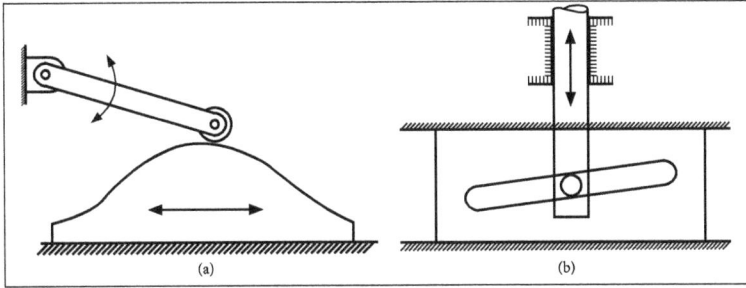

(a)   (b)

5.2 Pressure Angle and Under Cutting Base Circle Determination

The pressure angle, which is the reciprocal of the transmission angle m (i.e. a=p/2-m) is defined as:

$$\tan \alpha = \frac{\text{Force component tending to apply pressure on the follower bearings.}}{\text{Force component tending to move the follower.}}$$

In cams there is point contact between the two profiles. The force is transmitted along the common normal of the two contacting curves. Pressure angle for oscillating and translating roller follower radial cams are shown below. For flat faced followers the pressure is apparently zero at all times (the normal to the profile is normal to the flat face, which is always constant).

Curvature characteristics are used for the determination of cam size. In force closed cams, the pressure angle is important during the rise portion where cam is driving the follower, in return motion it is the spring force (or any other force used for forced closure) that lowers the follower.

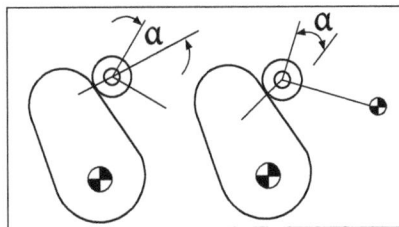

5.2.1 Forces and Surface Stresses

Forces

We can distinguish two important types of forces. These are the (distributed) contact forces t and the (distributed) mass forces b. They are defined as:

$$t = \lim_{\Delta A \to 0} \frac{\Delta F}{\Delta A} \quad \text{and} \quad b = \lim_{\Delta V \to 0} \frac{\Delta F}{\Delta V}.$$

Here F denotes a force, A denotes an area and V denotes a volume. Now let's examine a certain volume Ω. The total contact force F_s and the total body force F_b can be found using:

$$F_s = \int_{\partial \Omega} t(x) dA \quad \text{and} \quad F_b = \int_{\Omega} b(x) dV.$$

The signal $\partial \Omega$ means we integrate over the surface of the volume Ω. Together, the total contact force Fs and the total body force form the total external force $F_{ext.}$:

- Cast iron, on cast iron.
- Leather or metal.
- Cork on metal.
- Asbestos blocks on metal.
- Fiber on metal.

5.3   Design of Plate Clutches

When clutch engages, most of the work done (against friction forces opposing to motion) will be liberated as heat at the interface. Consequently the temperature of the rubbing surface will increase. This increased temperature may destroy clutch.

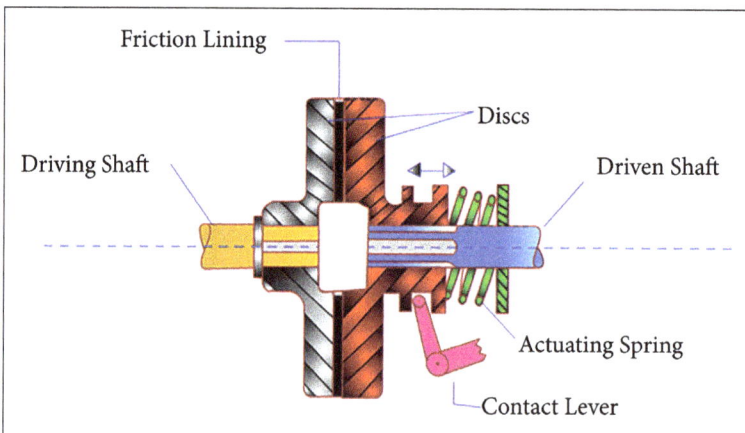

A single plate, friction clutch consisting of two flanges, as shown in figure above. One flange is rigidly keyed in to the driving shaft, while the other is free to move along the driven shaft due to spliced connection. The actuating force is provided by a spring, which forces the driven flange to move towards the driving flange. The face of the drive flange is linked with friction material such as cork, leather or foredo.

Types of Clutch

- Single plate clutch.

- Multi plate clutch.

- Cone clutch.

- Centrifugal clutch.

Uniform Pressure Theories

In uniform pressure theories total frictional torque acting on the friction surface or on the clutch is obtained by integrating the equation of the frictional torque on the elementary ring within limits from r_2 to r_1.

$$T = \mu WR$$

Where,

$$R = \frac{2}{3}\left[\frac{r_1^3 - r_2^3}{r_1^2 - r_2^2}\right]$$

Uniform Wear Theories

For uniform wear, the intensity of pressure varies inversely with the distance. Therefore,

p. r = C = constant.

$$T = \mu WR$$

Where, $k = \dfrac{r_1 + r_2}{2}$

Problems

1. A multi-disc clutch has three discs on the driving shaft and two on the driven shaft is to be designed for a machine tool, driven by an electric motor of 22 kw running at 1440 rpm. The inside diameter of the contact surface is 130 mm. The maximum pressure between the surfaces is limited 0.1 N/mm². Shall we discuss the design of a clutch? Take $\mu = 0.3$; m = 3; n2 = 2.

Given data:

Power = 22 Kw

N = 1440 rpm

$$d_2 = \text{Inside diameter} = 130 \text{ mm} \Rightarrow \frac{130}{2} = 65$$

Maximum pressure $(P_m ax) = 0.1 \text{ N/mm}^2$.

$\mu = 0.3$

$n_1 = 3$

$n_2 = 2$

Find $\gamma_1 =$ Outside diameter of contact surfaces.

$$P = \frac{2\pi N_s}{60}$$

$$T = \frac{60 \times P}{2\pi N} = \frac{60 \times 22 \times 10^3}{2\pi \, 1440} = \frac{1320000}{9043.2}$$

Solution

T = 145 N–m

$$T = 145 \times 10^3 \text{ N–mm}$$

$$P_{max}, \gamma_2 = C(\text{or})C = 0.1 \times \frac{130}{2} = 65 \text{ N/mm}$$

Assume uniform wear p.r = C, since the intensity of pressure is maximum at the inner radius (r_2).

We know that the axial force on each friction surface:

$$W = 2\pi c (r_1 - r_2)$$

$$= 2\pi \times 6.5 (r_1 - 65)$$

$$W = 40.82 \ (r_1 - 65)$$

For uniform wear, mean radius of the contact surface:

$$R = \frac{r_1 + r_2}{2} = \frac{r_1 + 65}{2} = 0.5\, r_1 + 32.5$$

Number of pairs of contact surfaces:

$$n = n_1 + n_2 - 1$$

$$= 3 + 2 - 1 = 4$$

$$n = 4$$

Torque transmitter $(T) = n.\mu.W - R$

$$145 \times 10^3 = 4 \times 0.3 \times 40.82 \left(\gamma_1 - 65\right)\left(\gamma_1 \times 0.5 + 32.5\right)$$

$$= 49 \left(\gamma_1 - 65\right)\left(0.5\gamma_1 + 32.5\right)$$

$$r_1 = 100 \text{ mm}$$

Result:

Outside diameter of disc $= r_1 = 100$ mm

2. A single plate clutch, both sides being effective, is required to connect a machine shaft to a driver shaft which runs at 500 rpm. The moment of inertia of the rotating parts of the machine is 1 kgm². The inner and outer radii of the friction discs are 50 mm and 100 mm respectively. Assuming uniform pressure of 0.1 N/mm and co-efficient of friction of 0.25, determine the time taken for the machine to reach full speed when the clutch is suddenly engaged. Also determine the power transmitted by the clutch, the energy dissipated during clutch slip and the energy supplied to the machine during engagement.

Given data:

$$N = 500 \text{ rpm}$$

$$M \cdot \text{Inertia} = 1 \text{ kg—m}^2$$

$$\text{Inner radii} = 50 \text{ mm}$$

$$\text{Outer radii} = 100 \text{ mm}$$

$$\text{Pressure} = 0.1 \text{ N/mm}$$

$$\text{Co-efficient of Friction} = 0.25$$

Solution

$$F = \pi p\left(R_o^2 - R_i^2\right)(\text{for uniform pressure}).$$

1. Axial Force:

$$= \pi \times 0.1 \left(100^2 - 50^2\right) = 2356.2 \text{ N}.$$

2. Torque Capacity:

$$T = F_\mu \cdot R_f \cdot n_p.$$

$$= F\mu \cdot \frac{2}{3}\left(\frac{R_o^3 - R_i^3}{R_o^2 - R_i^2}\right) n_p.$$

$$= 2356.2 \times 0.25 \times \frac{2}{3}\left(\frac{100^3 - 50^3}{100^2 - 50^2}\right) \times 2.$$

$$= 91630 \text{ N mm}.$$

3. Power:

$$\text{Power capacity of the clutch} = T_n\left(\frac{2\pi}{60}\right) \text{watts}.$$

$$= 91630 \times 10^{-3} \times 500 \times \frac{2\pi}{60}.$$

$$= 4795.3 \text{ watts}.$$

4. Time Taken for Acceleration:

The machine is accelerated to 500 rpm from zero speed. Torque required to accelerate, $T = I\alpha$,

Driver shaft torque = Mass MI of machine × angular acceleration

$$91.63 = 1 \times \alpha$$

$$\alpha = 91.63 \text{ rad/s}^2$$

We know, that $\omega = \omega_o + \alpha t$, ω_o = initial velocity = 0, and hence, $\omega = \alpha t$, where ω is the final angular velocity of the driven and t is the time taken to attain that velocity.

$$\frac{500 \times 2\pi}{60} = 91.63 \times 1$$

$$\therefore t = 0.571 \text{ sec}$$

5. Energy Dissipation:

Angle turned by the driver during slip.

During the time the driven is accelerated.

$$\theta_1 = \omega t = \frac{500 \times 2\pi}{60} \times 0.571 = 29.9 \text{ radians}$$

Driver runs at uniform angular velocity.

Angle turned by the driven:

$$\theta_2 = \omega_0 t + \frac{1}{2}\alpha t^2$$

$$\omega_0 = 0$$

$$\frac{1}{2}\alpha t^2 = \frac{1}{2} \times 91.63 \times 0.571^2$$

$$= 14.95 \text{ radians}$$

There is another way of finding θ_2.

Since, the machine accelerates from rest to 500 rpm, in 0.571 sec.

$$\theta_2 = \frac{1}{2}\left(\frac{500 \times 2\pi}{60}\right) \times 0.571 = 14.95 \text{ radians}$$

Energy lost in friction (dissipated as heat):

= work done by the motor−work done on the machine

$$= T(\theta_1 - \theta_2) = 91.63(29.9 - 14.95) = 1369.86 \text{ Nm}$$

Energy supplied to the machine:

$$= T\theta_2 = 91.63 \times 14.95 = 1369.9 \text{ Nm}$$

Alternatively,

Work done by the motor $= T\theta_1 = 91.63 \times 29.9 = 2739.7 \text{ Nm}$

K.E. acquired by the machine:

$$\frac{1}{2}I\omega^2 = \frac{1}{2} \times 1 \times \left(\frac{500 \times 2\pi}{60}\right)^2 = 1370.8 \text{ Nm}$$

Energy lost = 2379.7 − 1370.8 = 1369 Nm.

5.3.1 Axial Clutches

An axial clutch is one in which the mating frictional members are moved in a direction parallel to the shaft. A typical clutch is illustrated in the figure below. It consists of a driving disc connected to the drive shaft and a driven disc connected to the driven shaft.

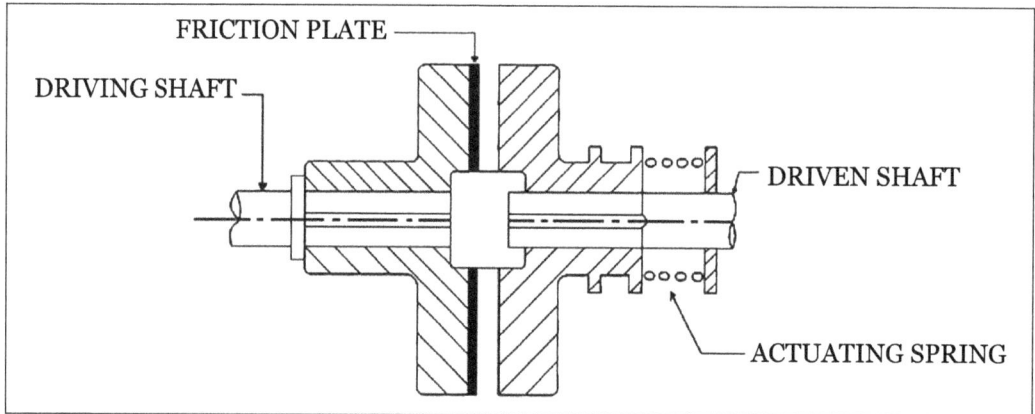

The friction plate is attached to one of the members. Actuating spring keeps both the members in contact and power/motion is transmitted from one member to the other. When the power of motion is to be interrupted the driven disc is moved axially creating a gap between the members as shown in the figure given above.

Torque Capacity - Depends:

- (T) Torque = F.N.R.n.

- F–Axial Force.

- N–Co-efficient of Friction.

- R–Friction Radius.

- n–Number of pairs of surfaces in contact.

- T – Torque (N-m).

- Clutch torque carrying capacity is depends upon the above mentioned factors.

5.3.2 Cone Clutches

The cone clutches are friction clutches. They are simple in construction and are easy to disengage. However, the driving and driven shafts must be perfectly coaxial for efficient functioning of the clutch. This requirement is more critical for cone clutch compared to single plate friction clutch. A cone clutch consists of two working surfaces, viz. inner and outer cones as shown in fig below.

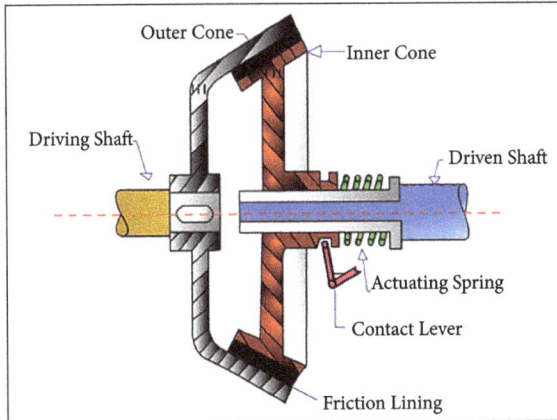

The output cone is fastened to the driving shaft and the inner cone is free to slide axially on the driven shaft due to splines. A spring provides the necessary axial force to the inner cone to press against the outer cone, thus engaging the clutch. A contact lever is used to disengage the clutch. The inner cone surface is lined with friction material.

The due to wedging action between the conical working surfaces, there is considerable normal pressure and friction force with a small engaging force. The semi cone angle a is kept greater than a certain value to avoid self-engagement; otherwise disengagement of clutch would be difficult. This is kept around 12.50.

Problems

A cone clutch is to transmit 7.5 kW at 900 rpm. The cone has a face angle of 12°. The width of the face is half of the mean radius ad the normal pressure between the contact faces is not to exceed 0.09 N/mm2. Assuming uniform wear and the coefficient of friction between contact faces as 0.2, let us determine the main dimensions of the clutch and the axial force required to engage the clutch.

Given data:

$a = 12°.$

$p = 7.5 \text{ kw}.$

$= 7.5 \times 10^3 \, \omega.$

$N = 900 \text{ rpm}.$

$p = 0.09 \times 10^{+6} \text{ N/m}^2.$

$\mu = 0.25.$

$S = \dfrac{R}{2} \; R = 26.$

If outer and inner diameter of the plane:

$$R = 28$$

We know that,

$$\frac{r_1 - r_2}{6} = \sin \alpha$$

$$S = \frac{r_1 - r_2}{\sin \alpha}$$

Mean Radius, $R = \frac{r_1 + r_2}{2}$

$$\frac{r_1 - r_2}{\sin \alpha} = \frac{r_1 + g_2}{2}$$

$$r_1 - r_2 = \frac{\sin 12°}{2}(r_1 + r_2)$$

$$r_1 - r_2 = 0.103(r_1 + r_2)$$

$$r_1 - r_2 = 0.103 \, r_1 + 0.103 \, r_2$$

$$r_1 - 0.013 \, r_1 = 0.103 \, r_2 + r_2$$

$$0.897 \, r_1 = 1.103 \, r_2$$

$$r_1 = 1.22 \, r_2$$

We know that power transmitted, $p = \frac{2\pi NT}{60}$

$$7.5 \times 10^3 = \frac{2 \times \pi \times 900 \times T}{60}$$

$$94.24 \, T = 7.5 \times 10^3$$

$$T = 79.58 \, N - m$$

Assuming services factor $k_s = 2.5$

Design Torque $[T] = T \times k_s$

$$= 79.58 \times 2.5$$

$$[T] = 198.95 \, N - m$$

Since, the intensity of the pressure is maximum, at the inner radius.

$$P_{max} \times r_2 = 0$$

Torque transmitted:

$$T = \mu \times \omega \times \operatorname{cosec} \alpha \left[\frac{r_1 + r_2}{2} \right]$$

$$\omega = 2\pi c (r_1 - r_2).$$
$$= 2\pi \, P_{max} \times \gamma_L (r_1 - r_2)$$

$$T = \mu \left[2\pi \, P_{max} \times r_2 (r_1 - r_2) \right] \operatorname{cosec} \alpha \left[\frac{r_1 + r_2}{2} \right]$$

$$[T] = \mu \times \pi \times P_{max} \times \operatorname{cosec} \alpha \times r_2 (r_1^2 - r_2^2)$$

$$198.95 = 0.25 \times \pi \times 0.09 \times 10^6 \times \operatorname{cosec} 12° \times r_2 (1.22 r_2)^2 - r_2^2$$

$$198.95 = 70685.83 \times 1.02 \times r_2 \left[1.48 r_2^2 - r_2^2 \right]$$

$$198.95 = 70685.83 \times 1.02 \times r_2 \left[0.48 \, r_2^2 \right]$$

$$198.95 = 34607.7 \, r_2^3$$

$$r_2^3 = \frac{198.95}{34607.7}$$
$$r_2^3 = 5.7 \times 10^{-3}$$

$$r_2 = 0.179 \text{ m}$$

$$r_2 = 179 \text{ mm}$$

$$\gamma_1 = 1.22 \times 179$$

$$\gamma_1 = 218.38 \text{ mm}$$

Axial force required:

$$\omega = 2\pi c (r_1 - r_2)$$

$$= 2\pi \, P_{max} \times \gamma_2 (r_1 - r_2)$$

$$= 2 \times \pi \times 0.09 \times 10^6 \times 179 \times \left(0.218.38 - 179\right)$$

$$= 2 \times \pi \times 0.09 \times 10^6 \times 0.179 \times \left(0.218 - 0.179\right)$$

$$= 3947.66 \text{ N.}$$

5.4 Internal Expanding Rim Clutches

An internal expanding brake is shown in fig below. It consists of a shoe, which is pivoted at 'A' and on the other end 'B' an actuating force F acts. A friction lining is provided on the shoe. The complete assembly of shoe with lining and pivot is placed inside the brake drum.

Under the action of the actuating force the shoe contact the inner surface of drum. Internal shoe brakes, with two symmetrical shoes are used in all automobile vehicles. The actuating force is usually provided by a hydraulic cylinder or a cam mechanism.

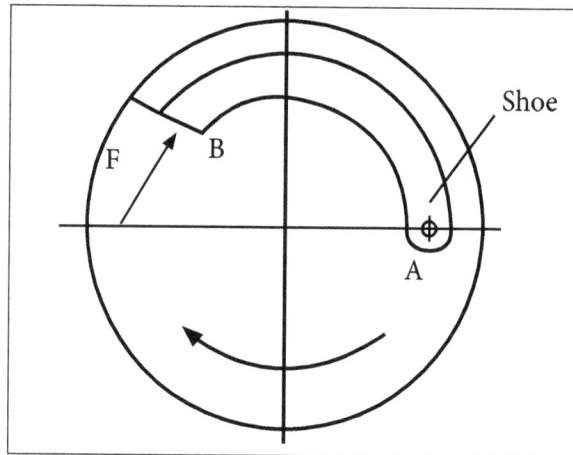

5.5 Electromagnetic Clutches

Electromagnetic clutches and brakes are electrically activated but transmit torque mechanically. Engagement time depends on magnetic field strength, air gap and inertia. Burnishing increases initial clutch or brake torque and over excitation cuts response time.

People use electromagnetic (EM) clutches and brakes every day and often don't realize it. Anyone who switches on a lawn tractor copy machine or car air conditioner may be using an EM clutch and EM brakes are just as common.

Electromagnetic clutches operate electrically but transmit torque mechanically. Engineers once referred to them as Electro mechanical clutches. Over the years EM came to stand for electromagnetic, referring to the way the units actuate but their basic operation has not changed.

Electromagnetic clutches and brakes come in many form, including tooth, multiple disc, hysteresis, and magnetic particle. However, the most widely used version is the single-face design.

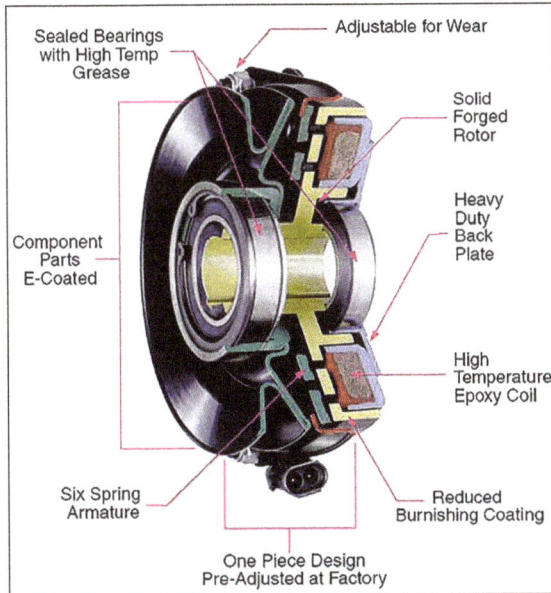

Elements of EM.

Both, EM clutches and brakes, share basic structural components, a coil in a shell, also referred to as a field a hub and an armature. A clutch also has a rotor, which connects to the moving part of the machine, such as a drive shaft.

5.6   Band and Block Brakes

In band brake, a flexible steel band lined with friction material, presses against the rotating brake drum. The braking action is performed either to slow down or halting the drum. The braking action is obtained by tightening the band around the drum. This kind of brake is used on sectional warping machine and warp let-off motion on conventional looms. Band brakes are classified into simple and differential band brakes.

A simple band brake is shown in Fig below, where one end of the steel band passes through the fulcrum of the actuating lever (O). The other end of the band is connected to the lever at point (A) a distance 'a' from the pivot point.

The band brake may be lined with blocks of wood or other material. The friction between the blocks and the drum provides braking action. Let there are 'n' number of blocks, each subtending an angle 2 θ at the centre and the drum rotates in anti-clockwise direction.

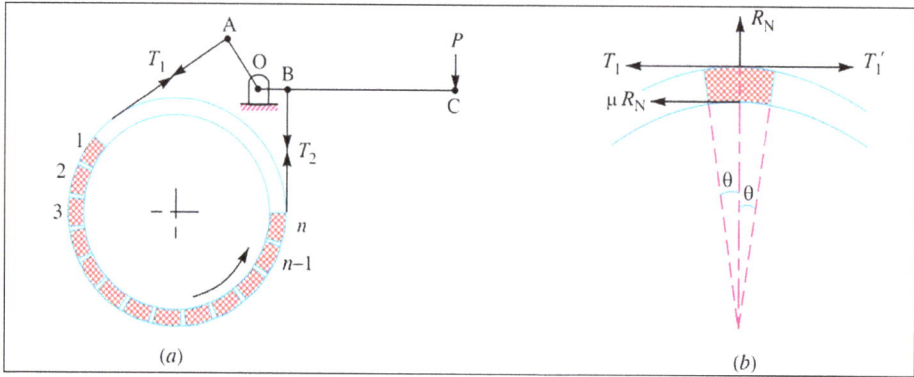

(a)   (b)

Let,

T₁ = Tension in the tight side.

T_1 = Tension in the tight side.

T_2 = Tension in the slack side.

μ = Coefficient of friction between the blocks and drum.

T_1' = Tension in the band between the first and second block.

T_2', T_3' etc. = Tension in the band between the second and third block, between the third and fourth block etc.

Consider one of the blocks (say first block). This is in equilibrium under the action of the following forces:

- Tension in the tight side (T1).

- Tension in the slack side (T1) or tension in the band between the first and second block.

- Normal reaction of the drum on the block (RN).

- The force of friction (μ.RN).Resolving the forces radially, we have:

$$\left(T_1 + T_1'\right)\sin\theta = R_N$$

Resolving the forces tangentially, we have:

$$\left(T_1 - T_1'\right)\cos\theta = \mu.R_N$$

Dividing above equations:

$$\frac{\left(T_1 - T_1'\right)\cos\theta}{\left(T_1 + T_1'\right)\sin\theta} = \frac{\mu.R_N}{R_N}.$$

$$\left(T_1 - T_1'\right) = \mu\tan\theta\left(T_1 + T_1'\right).$$

$$\frac{T_1}{T_1'} = \frac{1+\mu\tan\theta}{1-\mu\tan\theta}.$$

Similarly, it can be proved for each of the blocks that:

$$\frac{T_1'}{T_2'} = \frac{T_2'}{T_3'} = \frac{T_3'}{T_4'} = \dots = \frac{T_{n-1}}{T_2} = \frac{1+\mu\tan\theta}{1-\mu\tan\theta}.$$

$$\frac{T_1}{T_2} = \frac{T_1}{T_1'} \times \frac{T_1'}{T_2'} \times \frac{T_2'}{T_3'} \times \dots \times \frac{T_{n-1}}{T_2} = \left(\frac{1+\mu\tan\theta}{1-\mu\tan\theta}\right)^n.$$

Braking torque on the drum of effective radius r_e :

$$T_B = \left(T_1 - T_2\right)r_e.$$
$$= \left(T_1 - T_2\right)r.$$

5.6.1 External Shoe Brakes

In a shoe brake the rotating drum is brought in contact with the shoe by suitable force. The contacting surface of the shoe is coated with friction material. Different types of shoe brakes are used, viz., single shoe brake, double shoe brake, internal expanding brake, external expanding brake. These are sketched in fig below:

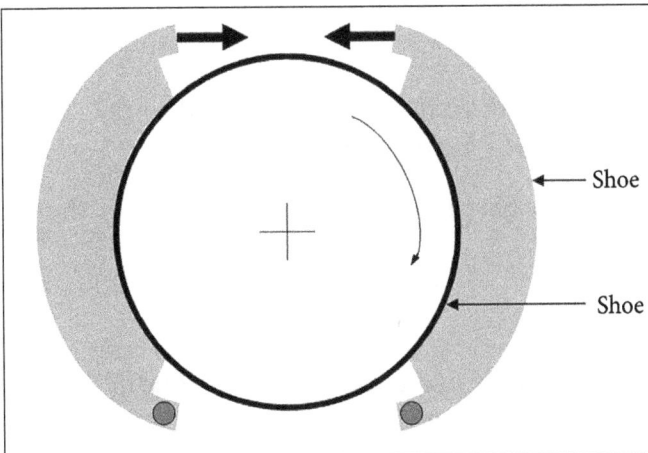

An external expanding shoe brake consists of two symmetrically placed shoes having inner surfaces coated with frictional lining. Each shoe can rotate about respective fulcrum. When the shoes are engaged, non-uniform pressure develops between the friction lining and the drum. The pressure is assumed to be proportional to wear which is in turn proportional to the perpendicular distance from pivoting point.

Self-Locking Brake

When the Frictional Force is great enough to apply the brake with no external force then the brake is said to be self-locking brake.

5.6.2 Internal Expanding Shoe Brake

An internal expanding brake is shown in Fig below. It consists of a shoe, which is pivoted at 'A' and on the other end 'B' an actuating force F acts. A friction lining is provided on the shoe. The complete assembly of shoe with lining and pivot is placed inside the brake drum. Under the action of the actuating force the shoe contact the inner surface of drum. Internal shoe brakes, with two symmetrical shoes are used in all automobile vehicles. The actuating force is usually provided by a hydraulic cylinder or a cam mechanism.

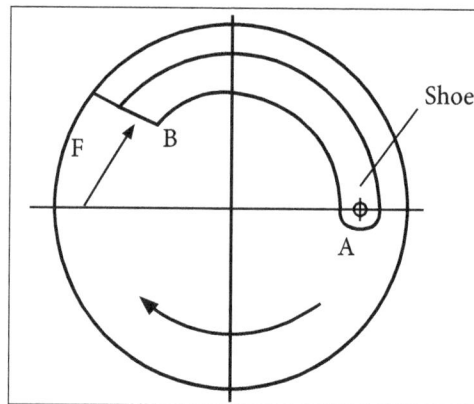

Self–Energizing Brake

When the frictional force is sufficient enough to apply the brake with no external force, then the brake is said to be self-locking brake.

When the frictional force helps in applying the brake, then the brake is said to be self-energized brake.

Problem

1. Let us determine the capacity and the main dimensions of a double block brake for the following data. The brake sheave is mounted on the drum shaft. The hoist with its load weights 45 kN and moves downwards with a velocity of 1.15 m/s. The pitch

diameter of the hoist drum is 1.25m. The hoist must be stopped with in a distance of 3.25 m. The kinetic energy of the drum may be neglected.

Solution

Given data:

Load = 45 Kn

Velocity = 1.15 m/s

Perimeter of drum = 1.25 m = 1250 mm

Hoist must stopped distance = 3.25 m

$$\text{Radius}(r) = \frac{1.25}{2} = 625 \text{ mm}$$

Find:

Design the double block.

$$\text{Mass} = \frac{450000}{10} = 45000 \text{ kg}$$

Solution

Kinetic energy of the mass:

$$E_k = \frac{1}{2}mV^2 = \frac{1}{2} \cdot 45000 \times (1.15)^2$$
$$= 29756.2 \text{ N}-\text{m}$$

We know that,

Potential Energy of the Mass (E_p):

$$E_p = m \cdot gh = 45000 \times 9.81 \times 3.25$$

$$E_p = 1434712.5 \text{ N}-\text{m}$$

Total Energy of the Moving mass (or) Energy to be absorbed by the brake:

$$E = E_k + E_p = 29756 + 1434712$$

$$E = 1464468 \text{ N}-\text{m}$$

Since, the mass is stopped as distance of 3.25 m; therefore, tangential braking force required is (F_t):

$$F_t \times distance = Total\ Energy\ (E)$$

$$F_t \times 3.25 = 1464468$$

$$F_t = 450605.5\ N$$

We know, that average braking torque to be applied to stop the mass is given by,

$$T_B = F_{Ext} = 450605 \times \left(\frac{1250}{2} \right)$$

$$T_B = 281628125\ N-m$$

Result:

Average braking torque = 281628125 N-m.

2. A rope drum of an elevator having 650 mm diameter is fitted with a brake drum of 1 m diameter. The brake drum is provided with four cast iron brake shoe each subtending an angle of 45°. The mass of the elevator when loaded is 2000 kg and moves with a speed of 2.5 m/s. The brake has a sufficient capacity to stop the elevator in 2.75 meters. Assuming the coefficient of friction between the brake drum and shoes as 0.2, find:

- Width of the shoe, if the allowable pressure on the brake shoe is limited to 0.3 N/mm².

- Heat generated in the stopping the elevator.

Given:

$$d_e = 650\ mm\ or\ r_e = 325\ mm\ ;\ d = 1\ m\ \ or\ \ r + 0.5\ m$$

$$n = 4\ ;\ 20 = 45°\ or\ \theta = 22.5°\ ;\ m = 2000\ kg\ ;\ V = 2.5\ m/s$$

$$h = 275\ m;\ \mu = 0.2;\ P_b = 0.3\ N/mm^2$$

Solution

(i) To find the width of the shoe (ω).

First, let us find the acceleration of the rope (a):

$$a = \frac{(2.5)^2}{5.5} = 1.136\ m/s^2$$

$$V^2 - u^2 = 2 \text{ ah or } (2.5)^2 - 0 = 2a(2.75) = 5.5 \text{ a}$$

Accelerating force, $m \times a = 2000 \, (1.136) = 2272$ N.

$\therefore$ Total load acting on the rope while moving,

W = Load on the elevator + Accelerating force.

$$(2000 \times 9.81) + 2272 = 21892 \text{ n}$$

Torque acting on the shaft,

$$T = \omega \times r_e = 21892 \times 0.325 = 7115 \text{ N} - \text{m}$$

$\therefore$ Tangential force acting on the drum:

$$= \frac{T}{r} = \frac{7115}{0.5} = 14230 \text{ N}$$

The brake drum is provided with four castor shoes, therefore tangential force acting on each shoe:

$$F_t = \frac{14230}{4} = 3357.5 \text{ N}$$

Since, angle of contact of each shoe is 45°, therefore the equivalent coefficient of friction (μ) can be calculated below:

$$R_N = \frac{F_t}{\mu} = \frac{3557.5}{0.2} = 17787.5 \text{ N}$$

The projected bearing area of each shoe,

$$A_b = \omega(2r \sin \theta) = \omega(2 \times 500 \sin 22.5°)$$

$$= 382.7 \, \omega \text{ mm}^2$$

Bearing pressure on the shoe (P_b):

$$0.3 = \frac{R_N}{A_b} = \frac{17787.5}{382.7 \, \omega} = \frac{46.5}{\omega}$$

$$\omega = \frac{46.5}{0.3} = 155 \text{ mm}$$

(ii) To find the heat generated in stopping the elevator:

Heat generated in stopping the elevator

= Total energy absorbed by the brake.

$$= \text{K.E.} + \text{P.E.} = \frac{1}{2}mv^2 + mgb$$

$$= \left[\frac{1}{2} \times 2000(2.5)^2\right] + \left[2000 \times 9.81 \times 2.75\right]$$

= 60205 N-m or 60.205 kJ.

Permissions

All chapters in this book are published with permission under the Creative Commons Attribution Share Alike License or equivalent. Every chapter published in this book has been scrutinized by our experts. Their significance has been extensively debated. The topics covered herein carry significant information for a comprehensive understanding. They may even be implemented as practical applications or may be referred to as a beginning point for further studies.

We would like to thank the editorial team for lending their expertise to make the book truly unique. They have played a crucial role in the development of this book. Without their invaluable contributions this book wouldn't have been possible. They have made vital efforts to compile up to date information on the varied aspects of this subject to make this book a valuable addition to the collection of many professionals and students.

This book was conceptualized with the vision of imparting up-to-date and integrated information in this field. To ensure the same, a matchless editorial board was set up. Every individual on the board went through rigorous rounds of assessment to prove their worth. After which they invested a large part of their time researching and compiling the most relevant data for our readers.

The editorial board has been involved in producing this book since its inception. They have spent rigorous hours researching and exploring the diverse topics which have resulted in the successful publishing of this book. They have passed on their knowledge of decades through this book. To expedite this challenging task, the publisher supported the team at every step. A small team of assistant editors was also appointed to further simplify the editing procedure and attain best results for the readers.

Apart from the editorial board, the designing team has also invested a significant amount of their time in understanding the subject and creating the most relevant covers. They scrutinized every image to scout for the most suitable representation of the subject and create an appropriate cover for the book.

The publishing team has been an ardent support to the editorial, designing and production team. Their endless efforts to recruit the best for this project, has resulted in the accomplishment of this book. They are a veteran in the field of academics and their pool of knowledge is as vast as their experience in printing. Their expertise and guidance has proved useful at every step. Their uncompromising quality standards have made this book an exceptional effort. Their encouragement from time to time has been an inspiration for everyone.

The publisher and the editorial board hope that this book will prove to be a valuable piece of knowledge for students, practitioners and scholars across the globe.

Index

www.ingramcontent.com/pod-product-compliance
Lightning Source LLC
Chambersburg PA
CBHW062003190326

41458CB00009B/2953